ISBN 978-1-333-57330-0
PIBN 10521498

SUPERSONIC FLOW AND SHOCK WAVES

A MANUAL ON THE MATHEMATICAL THEORY OF NON-LINEAR WAVE MOTION

APPLIED MATHEMATICS PANEL
NATIONAL DEFENSE RESEARCH COMMITTEE

CONFIDENTIAL

DISTRIBUTION LIST

AMP REPORT 38.2R

No. of Copies	
24	Office of Executive Secretary, OSRD
45	Liaison Office
	5 British Central Scientific Service
	5 British Admiralty Delegation
	2 P. M. S. Blackett
	1 Dr. Bronowski
	1 R. M. Davies
	1 R. P. Fraser
	1 S. Goldstein
	1 W. F. Hilton
	1 J. W. Maccoll
	1 W. Payman
	1 W. G. Penney
	1 W. C. F. Shepherd
	1 T. E. Stanton
	1 G. I. Taylor
	1 J. L. Tuck
	1 Col. P. Libessart
	1 M. S. Kisenko
	8 National Advisory Committee on Aeronautics
1	War Department Liaison Officer with NDRC Att: Capt. H. E. Clements
1	Coordinator of Research and Development, Navy Att: Lt. J. H. Wakelin
8	Aberdeen Proving Ground, Ordnance Research Center
	1 O. Veblen
	1 Capt. J. C. Clark
	1 R. H. Kent
	1 H. Lewy
	1 T. L. Smith
	1 J. Vinti
	1 R. J. Walker
3	Office of the Chief of Ordnance, Technical Service Division
	1 H. M. Morse
9	Chief, Bureau of Aeronautics, Navy Dept.
	1 Cdr. J. S. Warfel
	1 Lt. Cdr. W. Bollay
	1 E. S. Roberts
	5 Aerojet
3	Chief, Bureau of Ships, Navy Dept.
	1 Lt. Cdr. R. W. Goranson
	1 Lt. J. H. Curtiss
7	Director, David Taylor Model Basin
	1 Cdr. J. Ormondroyd
	1 G. H. Curl
	1 G. E. Hudson
	1 E. H. Kennard
8	Chief, Bureau of Ordnance, Navy Dept.
	1 Lt. Cdr. S. Brunauer
	1 R. S. Burington
	2 R. J. Seeger
	1 H. Polachek
	1 P. C. Keenan
7	R. C. Tolman, Vice Chairman, NDRC

No. of Copies	
10	E. B. Wilson, Chief, Divisio[n]
	4 Oceanographic Institutio[n]
	1 R. R. Halverson
	1 H. F. Bohnenblust
	1 R. Ladenburg
	1 F. Seitz
	1 L. Smith
	1 M. P. White
5	F. L. Hovde, Chief, Division 3
	1 C. N. Hickman
	1 C. C. Lauritsen
	1 F. T. McClure
	1 J. B. Rosser
1	H. B. Richmond, Chief, Divisio[n] Att: H. L. Dryden
6	J. T. Tate, Chief, Division 6
	1 W. V. Houston
	1 L. L. Foldy
	1 C. Herring
	1 C. L. Pekeris
	1 H. Primakoff
6	R. A. Connor, Chief, Division
	1 H. Eyring
	1 R. W. King
	1 D. P. MacDougall
	1 O. K. Rice
1	W. Weaver, Chief, Applied Math
	T. C. Fry
	R. Courant
	S. S. Wilks
	M. Rees
1	I. S. Sokolnikoff
1	G. D. Birkhoff
1	Garrett Birkhoff
1	J. G. Kirkwood
	P. M. Morse
1	J. v. Neumann
	W. Prager
	A. H. Taub
	H. Weyl

CONFIDENTIAL

PREFACE

The following book differs from the usual reports issued under the auspices of the Office of Scientific Research and Development. It does not answer specific questions, and it does not contain summaries of results which a casual reader could use without studying the analytical details; instead, basic theoretical aspects of gas dynamics are presented in a rather mathematical form for the increasing number of well-trained scientists in related war work.

While trying to make practical contributions to problems of gas flow, Dr. K. O. Friedrichs and the undersigned have found a thorough understanding of the theoretical background indispensable; the present manual is an attempt to condense the result of their efforts in this theoretical direction. Before being drawn into work for the NDRO, the writers were preparing a set of lecture notes on topics of mathematical physics. This plan had to be postponed, but contact with classified publications and active participation in work on the theory of explosions and gas dynamics made it possible to write this manual, covering far more ground than planned originally for a chapter in a volume on wave propagation.

The content of the following pages is largely conditioned by personal experience and interest. No attempt at a balanced presentation was made. Even less was it possible to appraise the merits of all recent contributors to the field. The names of scientists with whom the authors had much contact will appear frequently, while others are hardly mentioned. This is also true of the bibliography which, incidentally, contains a few references to sources of general information for those readers who find that the present book starts at a point somewhat beyond their general background.

The book was written during, not after, a period of study and investigation while points of view underwent gradual changes. As a consequence, for example, scant attention is given to phenomena in liquids as compared with those in gases. Important subjects, such as flow of compressible fluids around obstacles, are hardly touched. Under conditions of more leisure such shortcomings should have been remedied.

A collective effort of the New York University Group of the Applied Mathematics Panel was necessary to produce the book. In particular, the authors are indebted to Messrs. Charles DePrima, Harvey Cohn and John Knudsen. Much labor is embodied in the drawings (by Mr. John Knudsen), most of which represent relevant actual cases. But Mr. Robert Shaw, more than anyone else, has rendered invaluable help in all matters of technical, literary and scientific character. Without his help the book would contain far more errors than it does now. Even so, it is full of imperfections; its publication in the present state was prompted by the feeling that further delay might make it useless for the purpose for which it was planned.

For obvious reasons, detailed references to classified material and specific applications are not provided, and thus the table of contents may convey an impression of overemphasis on theoretical aspects. The following comments may, therefore, be made.

Chapter I contains classical facts underlying any mathematical treatment of compressible flow.

Chapter II develops a theory of the type of partial differential equations which occur in treatable problems of compressible flow. An important point is the emphasis on what the authors call "simple waves", representing motion in domains adjacent to domains of constant state.

Chapter III is a rather extensive analysis of motion in one dimension. After an initial mathematical discussion the basic types of continuous motion, so-called rarefaction waves, are studied. Then follows an analysis of discontinuous motion, that is, of shock waves. The last part deals with the phenomena that occur when such elementary motions interact, e.g., when shock waves and rarefaction waves collide with, or overtake, each other. Ultimately every motion of a gas must be analyzed by a study of such interactions. The theory of detonation waves, wave propagation in elastic-plastic solids under impact loading, and wave motion in open water channels are discussed briefly in the appendices to Chapter III.

Chapter IV deals with the case of steady two-dimensional flow, which presents itself most readily to a comprehensive mathematical analysis and, fortunately, provides an acceptable approximation to reality in many cases. Of particular interest to some readers may be the treatment of shock reflection, including the so-called Mach reflection.

Chapter V is of necessity the least systematic one. It deals with such problems in three dimensions as permit a reasonable theoretical attack. The first part concerns flow in nozzles and jets, a topic with increasingly important applications in many fields, e.g., rocket and jet propulsion. The second part is concerned with flow against conical obstacles such as projectiles, and gives an integrated summary of some work by Taylor and by Busemann. The problem of spherical waves, e.g., blast waves, is discussed very briefly in the last part of this chapter.

Altogether, the authors have tried to avoid discussions valuable mainly for their mathematical interest. Still, the book was written by mathematicians, and their willingness to accept compromise with an empirical approach does not make them physicists or engineers. Nevertheless, the authors hope that their effort will prove useful for the further development of the field.

R. Courant
Technical Representative
Contract OEMsr-945

C O N T E N T S

Page

III. MOTION IN ONE DIMENSION

A. Integration of the Differential Equations

B. Continuous Motion. Simple Waves

C. Shock Fronts

Page

I. INTRODUCTION

Violent disturbances - such as result from detonation of explosives, from flow through nozzles of rockets, from supersonic flight of projectiles, or from impact on solids - differ greatly from the "linear" phenomena of sound, light, or electromagnetic signals. In contrast to the latter, their propagation is governed by non-linear differential equations, and as a consequence, the familiar laws of superposition, reflection, and refraction cease to be valid; but even more novel features appear, among which the occurrence of shock fronts is the most conspicuous. Across shock fronts the medium undergoes sudden and often considerable increases in pressure and temperature. Even when the start of the motion is perfectly continuous, shock discontinuities may later arise automatically. Under other conditions, however, just the opposite happens; initial discontinuities may be smoothed out. Both these possibilities are essentially connected with the non-linearity of the underlying laws.

Nature confronts the observer with a wealth of non-linear wave phenomena, not only in flow of compressible fluids, but also in many other instances of practical interest. Such an example, rather different from those mentioned above, is the catastrophic pressure in a crowd of panicky people who rush toward a narrow exit or other obstruction. If they move at a speed exceeding that at which warnings are passed backward, a pressure wave arises much like that behind a shock front receding from a wall (see p. 77). In this manual, however, we shall refrain from further digressions into such fields and rather concentrate on the theory of compressible fluids.

Understanding and control of non-linear wave motion is a matter of obvious importance. Riemann, Earnshaw, Rankine, Hugoniot, Rayleigh and others wrote profound mathematical papers inaugurating this field of research almost a hundred years ago. Then the development was left mainly to a small group of ingenious men in the fields of mechanics and engineering. During the last few years, however, when the barriers between applied and pure science were forced down, there has arisen a wide-spread interest in non-linear wave motion, particularly in shock waves and expansion waves.

It is the purpose of the present manual to make the mathematical theory of non-linear waves more accessible, giving particular attention to recent developments.*

1. Qualitative differences between linear and non-linear waves. Some characteristics of non-linear wave motion can be described in general terms. In linear wave motion, such as in propagation of light or sound, local disturbances are always transmitted with a definite light speed or sound or sonic speed, which may vary within the medium and even in time, but which is a local property of the medium itself and remains the same for every conceivable wave motion in the medium. Such a sound speed also plays a rôle in non-linear wave motion. Small disturbances or "wavelets", slightly modifying a given primary wave motion, are propagated with a certain speed, again called sound speed, though in this case the sound speed depends not only on the medium but also on the specific character of the primary motion.

The distinctive feature of non-linear waves, however, concerns disturbances or discontinuities which are not necessarily small. In linear wave motion any initial discontinuity is preserved as a discontinuity and propagated with sonic speed. Non-linear wave motion behaves in a different manner. Suppose there is an initial discontinuity, e.g., between two regions of different pressures and flow velocities. Then we have the following alternative possibilities: either (1) such initial discontinuities are resolved immediately and the disturbance, while propagated, becomes continuous, or (2) the initial discontinuity will be propagated as a shock wave, a discontinuity advancing not at sonic but at

* As standard treatises on the theory of compressible fluid flow we refer to the following articles in the Bibliography at the end of the manual: Taylor and Maccoll [2], Busemann [3], Ackeret [4]. Extensive bibliographies are contained in these articles.

supersonic speed relative to the part of the medium into which the discontinuity penetrates. As previously stated, the shock fronts or shock waves are the most conspicuous phenomena occurring in non-linear wave propagation; they appear even without being caused by initial discontinuities. A continuous beginning does not ensure that the motion will subsequently remain continuous; after a while, discontinuities may develop automatically and be propagated further as shock waves.

In the mathematical theory, discontinuities which develop in this manner are represented by singularities of the solutions of the mathematical initial value problem. Unlike linear mathematical problems, non-linear ones often do not admit of solutions which can be continuously extended as far as the differential equations themselves remain regular.

The meaning of our alternative possibilities can be grasped more precisely if discontinuities are interpreted as idealizations or limits of continuous states. To fix our ideas, let us imagine a fluid medium in which at the beginning, i.e., at the time $t = 0$, the values of the velocities, the density ρ and the pressure p undergo jump discontinuities across the plane $x = 0$. Instead of attacking the corresponding initial value problem for the differential equation of gas dynamics directly, we consider the discontinuous initial values as limits of a sequence of continuous (even analytic) initial distributions. The initial value problem for the nth set of these approximating initial values certainly has a continuous and uniquely determined solution in a space-time neighborhood N_n of $x = 0$ and $t = 0$.

In the case of linear wave motion, the passage to the limit $n \to \infty$ for discontinuous initial values simply leads to the solution representing the propagation of the initial discontinuities at sound speed.* The non-linear case, however, exhibits a different behavior. To the nth of the approximating sets of continuous initial conditions there corresponds, as before, a neighborhood N_n in which the (continuous) solution of the initial value problem is determined (but beyond which possibly no continuous solution exists). Now the two alternative cases arise as follows. As n tends to infinity, or as the initial values approach the discontinuous distribution, either (1) all the neighborhoods N_n enclose a fixed neighborhood N of $t = 0$ and $x = 0$, in which case the solutions tend

* See Courant-Hilbert [12] , II, p:360.

in N to a continuous solution of the problem with discontinuous initial values; or (2) there is no such common neighborhood N, which means that the N_n shrink nearer and nearer to the place of the discontinuity. In the latter case the approximating solutions do not converge to a solution for the limit initial values, and our procedure in no way enables us to deduce a solution of the problem under consideration. Only further supplementary conditions, taken from physical facts, lead to a meaningful mathematical initial value problem. These general remarks will become more easily understood in the light of the specific details discussed later.

By the principle of superposition for linear waves, pressures of interfering sound waves are at most additive. In striking contrast to this fact interaction and reflection of non-linear waves may lead to enormous increases in pressure.

2. The medium.

(a) Gases and water. We shall be concerned with wave propagation in a medium whose state is described by quantities such as the density ρ or specific volume $\tau = \frac{1}{\rho}$, pressure p, entropy η per unit mass, internal energy e per unit mass, and temperature T. All these quantities may depend on the rectangular space coordinates x, y, z and the time t as well. Likewise, the particle velocity $\vec{q}$, with components u, v, w, in general also depends on x, y, z, t. We shall be concerned mainly with the motion of compressible fluids or gases, disregarding heat conduction and viscosity. Accordingly, for continuous motions at least, we assume that the particles of the medium undergo only adiabatic changes of state.

As known from thermodynamics, only two of the parameters ρ, p, T, η are independent; the specific thermodynamical nature of the medium is then characterized, for example, by a function defining T in terms of ρ and η and by a functional dependence

$$(1) \qquad p = f(\rho,\eta) \quad \text{or} \quad p = g(\tau,\eta) ,$$

which, with a certain freedom of nomenclature, we shall call the equation of state. Adiabatic changes correspond to η = constant along the path of a particle. Then, whenever the entropy η is initially the same throughout the medium, it retains its constant value during the motion and we are justified in simplifying (1) to a relation

$$(2) \qquad p = f(\rho) \quad \text{or} \quad p = g(\tau)$$

between the density, or the specific volume, and the pressure alone.

A basic assumption, well established by experience, is (for $\rho > 0$)

$$(3) \qquad f'(\rho) > 0 \quad \text{or} \quad g'(\tau) < 0 \ .$$

For most gases the equation of state is

$$(4) \qquad p = A\rho^{\gamma} \quad \text{or} \quad \frac{p}{p_0} = \left(\frac{\rho}{\rho_0}\right)^{\gamma} .$$

Such is the case for polytropic gases.* The factor A is related to the entropy η by

$$(5) \qquad \eta = \frac{1}{c_\tau} \log A \ ,$$

c_τ being the specific heat at constant density. γ is the adiabatic exponent, its value for air at normal temperature being $\gamma = 1.4$, and for all polytropic gases $\gamma > 1$. For liquids a similar equation of state,

$$(6) \qquad p = A\rho^{\gamma} - B \quad (\gamma = 7,\ B = 3000 \text{ atm. for water}),$$

where the pressure p is measured in atmospheres and A and B are independent of the entropy, has been empirically established as approximately valid for wide ranges. The qualitative behavior of polytropic gases will be seen to be dominated by the convexity of the function $p = g(\tau)$, expressed by

---

* For the definition see the footnote on page 12, Art. 4.

(7) $$g''(\tau) > 0.^{*}$$

(b) Elastic-plastic solids. (See also Appendix 2 to Chapter III). A quite different equation of state, and consequently a rather different type of non-linear wave motion, occurs in a solid slab capable of elastic and plastic deformations. Denoting by τ_0 the specific volume of the slab in the unstrained state and by ε the strain, we have

(8) $$\tau = \tau_0(1 + \varepsilon)$$

while the stress** σ is the negative of the pressure,

(9) $$\sigma = -p .$$

Instead of the equation of state (2) for gases, we now have to consider the stress-strain relation

(10) $$\sigma = \sigma(\varepsilon) = -g(\tau) ,$$

which, for elastic-plastic deformations, has the following character. With the elastic limit ε_* for the strain ε we have an elastic region $|\varepsilon| < \varepsilon_*$ where

$$\sigma = E\varepsilon = E\left(\frac{\tau}{\tau_0} - 1\right) ,$$

---

* Note that $f''(\rho) > 0$ implies (7) by virtue of (3).

** It appears from theory confirmed by experience that σ should be chosen as the "engineering stress", i.e., the force acting in normal direction on the cross-section of the slab divided by the original area of the cross-section of the unstrained slab.

E being Young's modulus. For $|\varepsilon| > \varepsilon_*$, i.e., beyond the elastic limit for the strain, the qualitative behavior of $g(\tau)$ is characterized by decreasing $\frac{d\sigma}{d\tau} = -g'(\tau)$ (in many actual cases decreasing to zero as τ approaches zero or infinity). For a graphical representation of the elastic-plastic stress-strain relation see Appendix 2 to Chapter III.

(c) Non-linear wave motion can also be studied in a medium consisting of a finite number of elements such as a chain of mass points connected with one another by elastic forces which obey a non-linear law of attraction and repulsion.

3. Differential equations of motion.* The phenomena to be studied in this manual will depend essentially on the general framework of the differential equations of hydrodynamics expressing Newton's law of motion and the principle of conservation of mass. By adding to these general statements the specific equation of state we obtain a complete system of differential equations which, together with appropriate initial and boundary conditions, determine an individual phenomenon.

In the following sections a brief survey of classical results is given in a form suitable for our purposes. The general equations of hydrodynamics can be expressed in two different forms, the form of Lagrange and the form of Euler. The equations in Lagrange's form describe the paths of the individual particles of the gas, i.e., the coordinates x, y, z of the particle, as functions of the time t and three parameters a, b, c which characterize the individual particle (a, b, c are often chosen as the coordinates of the particle at the time t = 0). In Lagrange's representation differentiation with respect to the time t will be denoted by a dot (˙) or by the symbol D.

In most cases, however, Euler's representation is preferable from a mathematical as well as from a physical point of view. This

---

* For this article see references Lamb [8] and Milne-Thomson [9] in the Bibliography.

representation aims at expressing more immediately observed quantities, namely, velocity, density, pressure, etc., as functions of the coordinates x, y, z and the time t. In Euler's representation, differentation with respect to the independent variables x, y, z, t is denoted by subscripts. The transition from Euler's representation to Lagrange's is effected by solving the system of ordinary differential equations

$$(11) \qquad \begin{cases} \dot{x} = u(x,y,z,t) \\ \dot{y} = v(x,y,z,t) \\ \dot{z} = w(x,y,z,t) \end{cases}$$

where a, b, c now appear as constants of integration.

Newton's law of conservation of momentum, supplemented by the statement of conservation of mass are formulated as the differential equations of motion. In Lagrangean form

$$(12) \qquad \begin{cases} \rho\ddot{x} + p_x = X \\ \rho\ddot{y} + p_y = Y \\ \rho\ddot{z} + p_z = Z \end{cases}$$

express the former, while

$$(13) \qquad (\rho\Delta)^{\cdot} = 0$$

expresses the latter, where $\Delta = \frac{\partial(x,y,z)}{\partial(a,b,c)}$ denotes the Jacobian of the functions x(a,b,c,t)', y(a,b,c,t), z(a,b,c,t) and where X, Y, Z, are the components of the external forces per unit mass, which we shall assume to be zero in this manual.

The derivatives of the pressure p in equations (12) refer to x, y, z, t as independent variables. An explicit expression in the variables a, b, c, t by $p_x = p_a a_x + p_b b_x + p_c c_x$, etc., will lead to involved non-linear terms, for a_x, b_x,... are to be expressed by the derivatives of the inverse functions $x(a,b,c,t),\ldots$. Usually, therefore, the Lagrangean representation becomes too cumbersome. This objection does not hold for one-dimensional motion, characterized by merely one space coordinate x; in this case the Lagrangean representation is often advantageous. For motion in more than one dimension, however, it is generally preferable to write the equations (without external forces) in Euler's form

$$\text{(14)}\qquad \begin{cases} u_t + uu_x + vu_y + wu_z + \frac{1}{\rho} p_x = 0 \\ v_t + uv_x + vv_y + wv_z + \frac{1}{\rho} p_y = 0 \\ w_t + uw_x + vw_y + ww_z + \frac{1}{\rho} p_z = 0 \\ \rho_t + u\rho_x + v\rho_y + w\rho_z + \rho(u_x + v_y + w_z) = 0 \end{cases} .$$

These four equations contain the five unknown quantities u, v, w, ρ, p and so constitute an "underdetermined" system of differential equations for these unknowns. In gas dynamics another equation, the equation of state (1), is added and an additional unknown, the entropy η, is introduced. Therefore one more relation is needed, and this missing equation is provided by the following crucial assumption, justified as long as we can disregard the effects of viscosity and heat conduction in the medium. the changes of state, as long as they remain continuous, are adiabatic; in other words, η does not change along the path of a particle. Thus, the four equations (14) are supplemented by the two equations

$$\text{(15)}\qquad p = f(\rho,\eta)$$

$$\text{(16)}\qquad \dot{\eta} = \eta_t + u\eta_x + v\eta_y + w\eta_z = 0 \quad .$$

Now the number of unknowns corresponds to the number of equations so that our system is "determined".

In most cases a further simplifying assumption can and will be made; viz., at the beginning of the process the entropy has the same value throughout the medium. In this case the flow will be called isentropic. Then, by (16), the entropy will retain its constant value, i.e., the flow will remain isentropic, and our system (14) is completed by the simpler relation

$$(17) \qquad p = f(\rho); \quad p_x = f'(\rho)\rho_x, \quad p_y = f'(\rho)\rho_y, \quad p_z = f'(\rho)\rho_z \,,$$

and we have four equations* for the four unknowns u, v, w, ρ. The quantity

$$c = \sqrt{f'(\rho)}$$

will play an important rôle throughout, and for reasons soon to become apparent, it is called the sound speed. Since $f'(\rho) > 0$, c is always real.

Of particular interest is the special case of a steady motion defined as a motion where the flow velocity, pressure, and density remain unchanged in time at each point, i.e., depend only on x, y, z and not on t. For such a motion the terms in our differential equations containing u_t, v_t, w_t, ρ_t and η_t drop out. In a steady flow all the particles passing through a particular point have the same velocity, density, pressure and entropy at this point and they will follow the same path, the streamline, through the point. The medium is thus covered by streamlines which do not change in time.

4. Remarks on the thermodynamical relations. Before completing the general framework of Euler's equations by specific relations between thermodynamical quantities, we recall briefly some underlying notions of the thermodynamics of a homogeneous medium. From the quantities T, η, $\rho = \frac{1}{\tau}$, p and the energy e two may be selected as independent variables.

---

* It is well known how the system for linear wave motion can be obtained from these equations by assuming that u, v, w, ρ deviate but little from normal constant values u_0, v_0, w_0, ρ_0. (See Courant-Hilbert [12], vol. II, pp. 305-6.)

Then the basic equation of thermodynamics, written in differentials, is

$$(18) \qquad de + p d\tau = T d\eta \quad .$$

If τ and η are considered as independent variables, the substance is thermodynamically characterized by the dependence of $e(\tau,\eta)$ on τ and η, and by (18) pressure and temperature are given as partial derivatives,

$$(19) \qquad e_\tau = -p, \quad e_\eta = T,$$

the first of the two relations expressing p in terms of τ or ρ with η as a parameter.

In gases, the pressure p depends noticeably on the entropy η. There are substances such as water, however, for which the pressure may be considered a function of density alone since the influence of changes in entropy are negligible. Then (18) shows that the temperature T depends only on η and that the energy e separates into the sum of a function of density alone and a function of entropy alone. Vice versa, such a <u>separation of the energy</u>, $e = e_1(\tau) + e_2(\eta)$, is also a sufficient condition for the dependence of p on ρ (or τ) alone.

Returning to the general case, we observe that it is often useful to introduce instead of the energy another quantity, known as the <u>enthalpy</u> or <u>heat content</u> i, given by

$$(20) \qquad i = e + \tau p \quad .$$

Then we can rewrite (18) in the form

$$(21) \qquad di - \tau dp = T d\eta \quad .$$

Introduction of the enthalpy proves particularly useful when adiabatic processes are considered, for then along a particle path the entropy η is a constant parameter and the enthalpy depends only on ρ or τ, so that along a particle path (21) simplifies to

(22) $$di = \tau \, dp$$

Therefore, for adiabatic changes we can write*

$$i = F(\rho)$$

with $$F'(\rho) = \frac{p'(\rho)}{\rho} = \frac{c^2}{\rho} \ ,$$

always with the understanding that $F(\rho)$ still depends on the entropy η as a parameter.

For constant entropy we may express the terms $\frac{1}{\rho}p_x, \frac{1}{\rho}p_y, \frac{1}{\rho}p_z$ in the equations (14), by means of the enthalpy i, as the components of the gradient of $i = F(\rho)$, this quantity considered as a function of x, y, z.

5. <u>Irrotational flow. Steady flow. Bernoulli's law.</u> Occasionally it is convenient to rewrite the differential equations (14) in vector notation. Under the assumption of isentropic flow, we have for the vector $\vec{q}$ of flow velocity and the enthalpy i,

(24) $$\dot{\vec{q}} + \operatorname{grad} i = 0 \ .$$

Rearranging terms,

(25) $$\vec{q}_t + \frac{1}{2}\operatorname{grad}(q^2) - \vec{q} \times \operatorname{curl} \vec{q} + \operatorname{grad} i = 0,$$

---

* This function is determined only within an additive function of the entropy, which for gases may be assumed to be a constant. Then $F(\rho)$ may be chosen as

(23) $$F(\rho) = \int_0^{\rho} \frac{dp}{\rho} = \int_0^{\rho} \frac{c^2}{\rho} d\rho$$

provided this integral exists for the particular equation of state in question. If in addition to this assumption relation (4), $p = A\rho^{\gamma}$ ($\gamma > 1$), holds, the gas will be called <u>polytropic</u>.

where q is the magnitude of the flow velocity or flow speed

$$q = |\vec{q}| = \sqrt{u^2 + v^2 + w^2}$$

and where the symbol × stands for vector multiplication.

Under rather wide assumptions, the equations of gas dynamics admit of important "integrals" which are easily deduced. (As always in this manual we restrict ourselves to the case where no external forces are acting).

We shall first formulate the law of conservation of circulation in isentropic flow. Let Γ be an arbitrary closed curve moving with the fluid. We consider the circulation C along Γ,

$$C = \oint_{\Gamma} udx + vdy + wdz \ ,$$

as a function of t. Then the theorem states that during the process the circulation remains constant: $\dot{C} = 0$. This follows almost immediately if we represent Γ by functions $x(\sigma,t), y(\sigma,t), z(\sigma,t)$, σ being a parameter on Γ such that Γ is described for $0 \leq \sigma \leq 2\pi$ and $u(0) = u(2\pi)$, etc. Then we have

$$\dot{C} = \oint_{\Gamma} (\dot{u}x_{\sigma} + \dot{v}y_{\sigma} + \dot{w}z_{\sigma} + u\dot{x}_{\sigma} + v\dot{y}_{\sigma} + w\dot{z}_{\sigma})\, d\sigma \ .$$

From the Lagrangean equations (12) and the form which (21) assumes when the entropy is constant, we have $\dot{x}_{\sigma} = u_{\sigma}, \ldots, \dot{u} = -\tau p_x = -i_x, \ldots,$ so that

$$(26) \qquad \dot{C} = \oint_{\Gamma} \left\{ \frac{1}{2} (u^2 + v^2 + w^2)_{\sigma} - i_{\sigma} \right\} d\sigma = 0 \ .$$

If the motion starts as an irrotational flow with curl $\vec{q} = 0$ (which is true for any motion starting from rest), then by Stokes' theorem we have, at the beginning, C = 0 for any curve Γ; and since by (26) C = 0 remains valid throughout the motion, the irrotational character of the motion is preserved, that is, curl $\vec{q} = 0$ holds throughout the motion. Consequently there exists a velocity potential, i.e., a function $\phi(x,y,z,t)$ such that

$$\vec{q} = \text{grad}\,\phi, \quad \text{or} \quad u = \phi_x, \quad v = \phi_y, \quad w = \phi_z \ .$$

The second consequence of the equations of motion is Bernoulli's law, which may be valid in the "weak" or "strong" sense depending on the assumptions made.

(a) Bernoulli's law for steady flows. A flow was called steady if local quantities such as ρ, u, v, w, p do not depend on the time t, so that in (25) the term q_t drops out. In steady flow the particles move along streamlines which are fixed in space and characterized by dx:dy:dz = u:v:w.

From the differential equations (24) we may immediately infer that on each streamline of a steady flow we have

$$0 = \vec{q}\,\dot{\vec{q}} + \vec{q}\ \text{grad}\ i = \frac{d}{dt}\left(\frac{1}{2}q^2 + i\right)$$

or

$$(27) \qquad \frac{1}{2}q^2 + i = \frac{1}{2}(u^2 + v^2 + w^2) + i = \frac{1}{2}\hat{q}^2 \ ,$$

where $\hat{q}$ is constant along a streamline (but need not necessarily have the same value along different streamlines). $\hat{q}$ will be called the ultimate or limit speed. Relation (27) is Bernoulli's law in the weak form. It is valid for steady, isentropic, but not necessarily irrotational, flow. If the limit speed $\hat{q}$ is the same for all streamlines, we speak of Bernoulli's law in the strong form.

(b) Bernoulli's law in the strong form is valid for flows which are irrotational and isentropic although not necessarily steady.

In terms of the velocity potential ϕ, Bernoulli's law is expressed as

$$(23)\qquad \frac{1}{2}\left(\phi_x^2 + \phi_y^2 + \phi_z^2\right) + \phi_t + i = \frac{1}{2}q^2 + \phi_t + i = \frac{1}{2}\hat{q}^2$$

where the limit speed $\hat{q}$ (which may depend on the time) is the same throughout the fluid, whether or not the flow is steady. The proof follows immediately from the form (25) of the equations of motion.

Along the streamline of a steady flow the pressure, density and flow speed vary. Bernoulli's equation $q^2 + 2i = \hat{q}^2$ establishes a one-to-one relation between q and i along the streamline of a steady flow. Since $\frac{di}{d\rho} = \frac{c^2}{\rho}$ is positive, the quantities ρ, $p = p(\rho)$, $c = \sqrt{p'(\rho)}$ are uniquely determined functions of q. We consider here q as an independent variable in the interval $0 \leq q \leq \hat{q}$; whether or not such values are actually assumed on the streamline under consideration is immaterial for this analytical aspect of Bernoulli's law. As q increases from zero on the difference $q^2 - c^2$ increases monotonically.* We now assume that $i = 0$, $c = 0$ when $\rho = 0$, as is true for polytropic gases; then for $\rho = 0$ the ultimate speed $\hat{q}$ is attained, and $q^2 - c^2$ is positive for $q = \hat{q}$. Since $q^2 - c^2$ is negative for $q = 0$, there exists a single intermediate value $c_* < \hat{q}$, called the <u>critical speed</u>, such that for $q = c_*$ flow speed and sound speed agree; $q = c = c_*$. For a given $\hat{q}$, wherever the flow speed is faster than the critical speed, $q > c_*$, it is automatically <u>supersonic</u>, i.e., $q > c$; and wherever $q < c_*$ the flow is <u>subsonic</u>, i.e., $q < c$. In other words, $q > c_*$ implies $q > c$ and $q < c_*$ implies $q < c$. This follows immediately from the fact that c decreases as q increases.

Evidently the value of the critical speed c_* (which we shall also occasionally denote by q_*) is determined by the value of $\hat{q}$ which characterizes the streamline in question.

This concept of a critical speed c_* separating the subsonic from the supersonic region does not depend on whether or not the critical speed is actually attained along a specific streamline in the flow. Every streamline has such a critical value associated with it in any case.

---

* This follows from $\frac{d}{dq^2}(q^2 - c^2) = 1 + \frac{1}{2}\frac{dc^2}{di} = 1 + \frac{1}{2}\frac{\frac{d}{d\tau}\frac{dp}{d\rho}}{\tau\frac{dp}{d\tau}} = 1 - \frac{1}{2}\frac{\frac{d}{d\tau}\tau^2\frac{dp}{d\tau}}{\tau\frac{dp}{d\tau}} = -\frac{1}{2}\frac{\tau\frac{d^2p}{d\tau^2}}{\frac{dp}{d\tau}} > 0$ by virtue of assumption (7).

Analytically we may characterize the critical speed for given $\hat{q}$ by first defining a critical value ρ_*. Setting $q^2 = c^2 = p' = f'(\rho)$ in (27), we obtain

$$p'(\rho_*) + 2i(\rho_*) = \hat{q}^2 .$$

This relation then determines ρ_* and hence $c_*^2 = p'(\rho_*)$ uniquely.

In the case of polytropic gases with the equation of state $p = A\rho^\gamma$ (see equation (4), Art.2(a)), the form of Bernoulli's law is particularly simple. We easily obtain

$$i = A\frac{\gamma}{\gamma - 1}\rho^{\gamma - 1} = \frac{\gamma}{\gamma - 1}\frac{p}{\rho}$$

and

$$c^2 = \gamma\frac{p}{\rho} ,$$

where A still depends on the entropy. Bernoulli's law for steady flow becomes

$$\frac{1}{2}q^2 + \frac{\gamma}{\gamma - 1}\frac{p}{\rho} = \frac{1}{2}\hat{q}^2 , \text{ or}$$

(29)
$$q^2 + \frac{2}{\gamma - 1}c^2 = \hat{q}^2 ,$$

and the critical speed c_*, obtained for $q = c = c_*$, is expressed by

(30)
$$c_*^2 = \frac{\gamma - 1}{\gamma + 1}\hat{q}^2 .$$

Setting

$$\frac{\gamma - 1}{\gamma + 1} = \mu^2$$

we have

(31)
$$c_* = \mu\hat{q} ,$$

and we may write Bernoulli's law in the form

$$\mu^2 q^2 + (1 - \mu^2) c^2 = c_*^2 \tag{32}$$

or

$$c = c_* \sqrt{\frac{1 - \mu^2 \left(\frac{q}{c_*}\right)^2}{1 - \mu^2}} \tag{32'}$$

6. Various forms of the differential equations. For steady isentropic irrotational flow we obtain a simple differential equation of second order for the velocity potential $\phi(x,y,z)$. First we use the three equations of motion in (14) to express the quantities $\frac{\rho_x}{\rho}$, $\frac{\rho_y}{\rho}$, $\frac{\rho_z}{\rho}$ in terms of $c^2 = p'(\rho)$ and u, v, w and their derivatives. Substituting in the equation of conservation of mass in (14), we find

$$\begin{aligned} (c^2 - u^2)\phi_{xx} + (c^2 - v^2)\phi_{yy} + (c^2 - w^2)\phi_{zz} \\ - 2uv\phi_{xy} - 2vw\phi_{yz} - 2wu\phi_{zx} = 0 \end{aligned} \tag{33}$$

with c^2 given as a function of $q^2 = u^2 + v^2 + w^2$ by Bernoulli's law (32). Now $u = \phi_x$, $v = \phi_y$, $w = \phi_z$, so that (33) represents a single differential equation for one unknown function $\phi(x,y,z)$.

In the special case of steady flow in two dimensions we have $w = 0$, all the derivatives with respect to z vanish and the original differential equations can easily be replaced by the following system of the first order

$$\begin{cases} u_y - v_x = 0 \\ u_x(c^2 - u^2) - (u_y + v_x)uv + v_y(c^2 - v^2) = 0 \end{cases} \tag{34}$$

with c^2 defined by Bernoulli's equation (32). This system is equivalent to the original equations with the added assumption that Bernoulli's law holds in the strong sense or that the flow is irrotational. Hence the equivalence is true, in particular, for flows that start from rest.

Another and even more simple form of the differential equations, immediately expressing the irrotational character of the flow and the principle of conservation of mass, is

$$(35) \qquad \begin{cases} u_y - v_x = 0 \\ (\rho u)_x + (\rho v)_y = 0 \end{cases}$$

which, of course, is equivalent to the preceding form by Bernoulli's law.

7. Lagrange's equations of motion for one-dimensional flow. In the case of one-dimensional motion, i.e., motion for which every quantity depends on the time and on only one space variable x, the existence of a velocity potential ϕ is trivial and not especially interesting. In this case, however, it is often convenient to use the differential equations in the Lagrangean form, which become particularly simple upon introducing in place of the density ρ the mass h in a column of unit cross-section bounded by planes through an arbitrarily chosen "zero" particle with the coordinate x_0 and the particle with the coordinate x (both of which move in time). We then consider x and ρ as functions of h and t. We have

$$\int_{x_0}^{x} \rho dx = h \quad ,$$

whence

$$\frac{1}{\rho} = \tau = x_h \quad .$$

Moreover, since $-p_x$ is the force per unit volume, the force per unit mass is $-\tau p_x = -p_h$. Thus Newton's law is expressed by

$$x_{tt} = -p_h \quad . \tag{36}$$

The law of conservation of mass is already implicit in the choice of our variables, while the assumption of adiabatic changes will now in general have the form $\eta = \eta(h)$. If, in particular, the entropy is constant throughout, we have with the equation of state $p = g(\tau)$ the relation

$$p_h = g'(\tau)\tau_h \tag{37}$$

and hence by (36) we obtain

$$x_{tt} = k^2(x_h)x_{hh} \tag{38}$$

where

$$k(\tau) = \sqrt{-g'(\tau)} = \rho c \quad .$$

$k = \rho c$ is often called the <u>impedance</u> of the medium. Equation (38) is the Lagrangean form of the equation of motion as a single differential equation of second order. This form seems somewhat more convenient than the form referring directly to the initial coordinate a of the particle x, with which the mass h is connected by $h = a\rho_0$, where ρ_0 is a constant density at time $t = 0$ and $a = 0$ corresponds to x_0.

Instead of this single differential equation of second order we may write the equivalent <u>system of first order</u> by considering the two unknowns $x_h = \tau = \frac{1}{\rho}$ and $u = x_t$ satisfying the system

$$\begin{cases} u_h = \tau_t \\ u_t = k^2(\tau)\tau_h \end{cases} \quad . \tag{39}$$

II. MATHEMATICAL THEORY OF ISENTROPIC FLOWS DEPENDING ON TWO INDEPENDENT VARIABLES

8. The differential equations. The general case of three-dimensional non-steady flows is much too involved for analytic treatment except when only two independent variables occur and when the flow is assumed to be isentropic. This is typical of the following cases:

(a) One-dimensional fluid motion, i.e., motion in which the state depends only on the time t and on one space coordinate x; for example, gas in a cylindrical tube along the x-axis, the independent variables being the distance x and the time t.

(b_1) Axially symmetric two-dimensional motion, where the state depends only on the time t and on the distance x from a fixed axis normal to the plane of motion.

(b_2) Three-dimensional flow with spherical symmetry about the origin, the independent variables being x and t, where x is now the radial distance from the origin.

(c) Steady flow in the x,y-plane.

(d) Steady flow in space with symmetry about the x-axis, the independent variables being x and the distance y from the x-axis.

The following are the differential equations corresponding to these various cases:

(a) Flow in one dimension. The flow at any point x at time t is characterized by the velocity u, the pressure p and the density $\rho = \frac{1}{\tau}$. If we assume that $p = p(\rho)$ and $\frac{dp}{d\rho} > 0$, and introduce the sound speed $c = \sqrt{\frac{dp}{d\rho}}$, the equations of flow become

$$\text{(A)} \qquad \begin{cases} u_t + uu_x + \frac{c^2}{\rho}\rho_x = 0 \\ \rho_t + \rho u_x + u\rho_x = 0 \end{cases}$$

or, for the dependent variables u and the enthalpy i (I(23), Art. 5) (ρ and

c^2 being considered as functions of i),

$$\text{(B)} \qquad \begin{cases} u_t + uu_x + i_x = 0 \\ i_t + c^2 u_x + ui_x = 0 \;. \end{cases}$$

For a polytropic gas (or water) with the adiabatic exponent γ, we may introduce u and c as dependent variables; the equations then become

$$\text{(C)} \qquad \begin{cases} u_t + uu_x + \dfrac{2}{\gamma - 1}\, cc_x = 0 \\ c_t + \dfrac{\gamma - 1}{2}\, cu_x + uc_x = 0 \end{cases}$$

In Lagrangean form, with the independent variables t and $h = \int_{x_0}^{x} \rho(\xi, t)\, d\xi$ and the dependent variables $x_h = \tau$ and $u = x_t$, we have the system

$$\text{(D)} \qquad \begin{cases} u_h = \tau_t \\ u_t = k^2(\tau)\tau_h \end{cases}$$

where $k^2(\tau) = c^2(\rho)\rho^2$.

(b_1) Two-dimensional flows with axial symmetry. With x as the radial distance from the axis of symmetry and u as the radial velocity the equations are

$$\text{(E)} \qquad \begin{cases} u_t + uu_x + \dfrac{c^2}{\rho}\rho_x = 0 \\ \rho_t + \rho u_x + u\rho_x + \dfrac{\rho u}{x} = 0 \;. \end{cases}$$

(b_2) Three-dimensional flow with spherical symmetry about the origin. Here, with x as the radial distance from the origin and u as the radial velocity, the equations become

$$
\text{(F)} \qquad \begin{cases} u_t + uu_x + \dfrac{c^2}{\rho}\rho_x = 0 \\[2ex] \rho_t + \rho u_x + u\rho_x + \dfrac{2\rho u}{x} = 0 \quad . \end{cases}
$$

The essential difference between equations (E) and (F) and those of case (a) is the presence of the terms $\frac{\rho u}{x}$ and $\frac{2\rho u}{x}$, respectively, in which the independent variable occurs explicitly.

(c) Steady flow in two dimensions. Here x and y are the independent variables and the velocity components u and v are the dependent variables. We assume that Bernoulli's law in the strong sense applies, which for isentropic flow is tantamount to the assumption of irrotational flow. Then c is a function of $q^2 = u^2 + v^2$, this relationship being given in the case of a polytropic gas by

$$
c^2 = \frac{\gamma - 1}{2}(\hat{q}^2 - q^2) \qquad \hat{q} = \text{constant.}
$$

In this case the equations of motion become (see equations (34), Art.6)

$$
\text{(G)} \qquad \begin{cases} u_y - v_x = 0 \\[2ex] u_x(c^2 - u^2) - (u_y + v_x)uv + v_y(c^2 - v^2) = 0 \quad . \end{cases}
$$

(d) Steady flow with symmetry about the x-axis. Again assuming Bernoulli's law in the strong sense and considering as the dependent variables the axial component of the velocity u and the radial component of velocity v, with x and the radial distance y from the x-axis as the independent variables, we have as the equations of motion

$$
\text{(H)} \qquad \begin{cases} u_y - v_x = 0 \\[2ex] u_x(c^2 - u^2) - (u_y + v_x)uv + v_y(c^2 - v^2) + \dfrac{c^2 v}{y} = 0 \quad . \end{cases}
$$

As in the case of three-dimensional flow with spherical symmetry and two-dimensional flow with axial symmetry, the essential difference between this flow and steady plane flow is the occurrence of a term $\frac{c^2 v}{y}$ containing explicitly the independent variable y which measures the radial distance.

All these systems have a similar mathematical structure; they consist of two quasi-linear partial differential equations of first order in two dependent and two independent variables. A general mathematical theory applicable to all these types of flow will clarify and unify the analysis of the various special cases.

Let us denote by u, v the dependent and by x, y the independent variables. These will later be identified with the various dependent and independent variables of the preceding problems. Then the general form of the differential equations is

$$(1) \quad \begin{cases} A_1 u_x + B_1 u_y + C_1 v_x + D_1 v_y + E_1 = 0 \\ A_2 u_x + B_2 u_y + C_2 v_x + D_2 v_y + E_2 = 0 \end{cases}$$

where the A,B,C,D,E are known functions of x,y,u,v. If $E_1 = E_2 = 0$ the system is homogeneous. If the A,B,C,D,E are functions of x,y alone the equations are linear and are consequently much easier to handle; and if A,B,C,D are functions of u,v alone and $E_1 = E_2 = 0$, a similar simplification will present itself.

In this case the differential equations (1) are made linear by interchanging the rôles of dependent and independent variables. For, if x and y are considered functions of u and v, we have

$$u_x = \Delta y_v, \qquad u_y = -\Delta x_v,$$

$$v_x = -\Delta y_u, \qquad v_y = \Delta x_u,$$

provided that the Jacobian

$$\Delta = u_x v_y - u_y v_x$$

does not vanish. Hence, in our special case, the equations (1) are transformed into the linear differential equations

$$(1') \qquad \begin{cases} A_1 y_v - B_1 x_v - C_1 y_u + D_1 x_u = 0 \\ A_2 y_v - B_2 x_v - C_2 y_u + D_2 x_u = 0 \end{cases} .$$

This situation occurs in the case (a) of one-dimensional general flow and the case (c) of two-dimensional steady flow.

The possibility of this linearization depends essentially on the assumption $\Delta \neq 0$, and hence those solutions for which $\Delta = 0$ are excluded. We shall see, however, that these latter are of particular interest.

9. Characteristic parameters. The theory depends upon the transformation of system (1) to a normal form or characteristic form by the introduction of two suitable new independent variables α, β (so-called characteristic parameters) in place of x and y. In this characteristic form the two equations (1) for u and v as dependent, and x and y as independent variables are replaced by four equations for the four quantities u, v, x, y as dependent variables and the two parameters α, β as independent variables. Thus a decided simplification can be attained; in two of the new equations only differentiations with respect to α, and in the remaining two only differentiations with respect to β will occur, and the independent variables α, β do not occur explicitly.

The derivation can be indicated briefly. Let us consider a solution $u(x,y)$, $v(x,y)$ of (1) and, in particular, the values of u and v along a curve L given by a parameter σ in the form $x = x(\sigma)$,

$y = y(\sigma)$. Along L, u and v also become functions of σ and, if we denote differentiation with respect to σ by a dot, we have along L

$$(1^*) \quad \begin{cases} \dot{x}u_x + \dot{y}u_y - \dot{u} = 0 \\ \dot{x}v_x + \dot{y}v_y - \dot{v} = 0 \end{cases}$$

If $x(\sigma)$, $y(\sigma)$, $u(\sigma)$, $v(\sigma)$ are considered as given, then (1) and (1^*) form a system of four linear equations for the four quantities u_x, u_y, v_x, v_y along L. In general, therefore, the given quantities $x(\sigma)$, $y(\sigma)$, $u(\sigma)$, $v(\sigma)$ will determine uniquely the derivatives u_x, u_y, v_x, v_y along L. It is natural, however, to investigate the exceptional case when the determinant of the coefficients, i.e., the first determinant of the matrix

$$\begin{pmatrix} A_1 & B_1 & C_1 & D_1 & E_1 \\ A_2 & B_2 & C_2 & D_2 & E_2 \\ \dot{x} & \dot{y} & 0 & 0 & -\dot{u} \\ 0 & 0 & \dot{x} & \dot{y} & -\dot{v} \end{pmatrix}$$

vanishes:

$$(A_1C_2 - A_2C_1)\dot{y}^2 - (A_1D_2 - A_2D_1 + B_1C_2 - B_2C_1)\dot{x}\dot{y} + (B_1D_2 - B_2D_1)\dot{x}^2 = 0.$$

In this exceptional case, since the system (1), (1^*) according to our assumptions also has a solution, not only the determinant vanishes, but in addition the whole matrix has rank not more than three, which yields one more independent relation.

Now we consider the two values ζ_+ and ζ_- of the ratio $\dot{y} : \dot{x}$ for which the determinant is zero; if ζ_+ and ζ_- are real and different, then we have for our solution u(x,y), v(x,y) two different families of such exceptional curves L. Calling our parameter σ either α or β according to which of the two roots ζ we choose, we obtain the transformation described in the text above. It is easily seen that if this procedure is reversed, the four equations in the text lead by elimination of α and β to the differential equations (1) and thus are completely equivalent to them.

The characteristic differential equations equivalent to (1) can be written in the following form:

$$(2)\quad \begin{cases} I_+ & y_\alpha - \zeta_+ x_\alpha = 0 \\ I_- & y_\beta - \zeta_- x_\beta = 0 \\ II_+ & u_\alpha + (R\zeta_+ - S)v_\alpha + (K\zeta_+ - H)x_\alpha = 0 \\ II_- & u_\beta + (R\zeta_- - S)v_\beta + (K\zeta_- - H)x_\beta = 0 \end{cases}$$

where ζ_+ and ζ_- are defined as the two roots of the quadratic equation

$$(3)\qquad [AC]\zeta^2 - ([AD] + [BC])\zeta + [BD] = 0 ,$$

and where

$$R = \frac{[AC]}{[AB]}, \qquad S = \frac{[BC]}{[AB]},$$

$$K = \frac{[AE]}{[AB]}, \qquad H = \frac{[BE]}{[AB]},$$

the brackets denoting the determinant

$$[XY] = \begin{vmatrix} X_1 & Y_1 \\ X_2 & Y_2 \end{vmatrix}$$

The possibility of the characteristic transformation of (1) given above depends on the <u>hyperbolic character</u> of the system, expressed by the inequality

$$(4)\qquad ([AD] + [BC])^2 - 4[AC][BD] > 0$$

This condition is satisfied for all our problems, with the qualification that in the case of steady flow the velocity is supersonic, i.e., $q^2 = u^2 + v^2 > c_*^2$. Furthermore, it is supposed that the conditions $[AB] \neq 0$, and therefore $[CD] \neq 0$, as well as $[AC] \neq 0$ and $[BD] \neq 0$

are satisfied in all cases under consideration.*

Under these conditions the quadratic equation (2) yields two different real solutions ζ_+, ζ_- (different from zero and infinity) as functions of u,v,x,y.

Obviously, without changing the form of the characteristic equations, we may replace α by $\alpha' = W(\alpha)$ and β by $\beta' = V(\beta)$ where W and V are arbitrary monotonic functions.

The main point is that under our assumption the system (1) is equivalent to our characteristic system in the sense that a solution of either yields a solution of the other. To this we add a few further remarks and conclusions.

For a specific solution of our system, we consider the two families of lines α = constant and β = constant in the x,y-plane and their images in the u,v-plane, and call them the characteristics C and the characteristics Γ respectively, distinguishing

$$C_+, \Gamma_+ \ : \ \beta = \text{constant},$$

$$C_-, \Gamma_- \ : \ \alpha = \text{constant}.$$

Two special cases of our system (1) are particularly simple:

(1) When the differential equations (1) are <u>linear</u>, i.e., when $A_1, \ldots, E_2$ depend on x and y only, then by (3) ζ_+ and ζ_- are known functions of x and y alone and the differential equations I are not coupled with the equations II. The equations I are equivalent to two ordinary differential equations

* If [AC] = [BD] = 0, the equations (1) could be put into the desired normal form by a simple elimination. Furthermore, if one of the determinants [AC], [CD] vanishes, a rotation of the coordinate system will correct this. [AB] and [CD] are assumed different from zero for obvious reasons.

(5)
$$\begin{cases} C_+: & \frac{dy}{dx} = \zeta_+(x,y) \\ C_-: & \frac{dy}{dx} = \zeta_-(x,y) , \end{cases}$$

which determine the families C_+ and C_- respectively. In the linear case the C characteristics in the x,y-plane are fixed, i.e., they do not depend on the specific solutions u,v of (1).

(2) If $E_1 = E_2 = 0$, and if $A_1,\ldots,D_2$ depend on u,v only (as in the case of steady supersonic flow in two dimensions) the situation is similar. Then ζ_+ and ζ_- are known functions of u and v and the differential equations II are independent of x and y.* They determine the characteristics Γ in the u,v-plane, or what is equivalent, the ordinary differential equations

(6)
$$\begin{cases} \Gamma_+ : & \frac{du}{dv} = S - R\zeta_+ \\ \Gamma_- : & \frac{du}{dv} = S - R\zeta_- \end{cases}$$

determine the families Γ_+ and Γ_- respectively. (For the underlying reason behind the reciprocity of these two special cases, see the remark at the end of Art.8).

10. Initial value problem. Domain of Dependence. Range of Influence. In the preceding article the characteristics appeared essentially as a mathematical tool for simplifying the form of the differential equations. Physically, however, the point of primary interest is the rôle of the characteristics as lines of propagation of disturbances, or Mach lines. This rôle together with the outstanding mathematical features of wave propagation in general are

---

* The same, incidentally, remains true even if E_1 and E_2 still depend on u and v.

best understood in connection with the basic problem of hyperbolic differential equations, the initial value problem, which we consider for system (1). Given a curve L in the x,y-plane by means of a parametric representation $x = x(\sigma)$, $y = y(\sigma)$ and along L values $u(\sigma)$, $v(\sigma)$ such that L is nowhere characteristic or tangent to a characteristic; the problem then is to find a solution of (1) which assumes the values $u(\sigma)$, $v(\sigma)$ on L.

To solve the problem* we apply the transformation to characteristic parameters α and β, which, with no loss of generality, may be normed so that the initial curve Λ in the α,β-plane is represented by Λ: $\alpha + \beta = 0$.

The initial value problem can now be formulated for the differential equations I, II in the α,β-plane. On the line Λ the values of x,y,u,v are prescribed, and in a neighborhood of Λ a solution of the characteristic equations I and II is sought which assumes these given values on Λ.

To construct the solution we differentiate I_+ and II_+ with respect to β and I_- and II_- with respect to α and thus obtain four linear equations in $x_{\alpha\beta}$, $y_{\alpha\beta}$, $u_{\alpha\beta}$, $v_{\alpha\beta}$. The determinant of these linear equations has the value $(\zeta_+ - \zeta_-)^2R$ which is different from zero. Hence we can solve for $x_{\alpha\beta}$, $y_{\alpha\beta}$, $u_{\alpha\beta}$, $v_{\alpha\beta}$ and obtain a system of equations of the form

$$(7) \qquad x_{\alpha\beta} = f_1, \quad y_{\alpha\beta} = f_2, \quad u_{\alpha\beta} = f_3, \quad v_{\alpha\beta} = f_4,$$

where the functions f depend on all the quantities $x,y,u,v,x_\alpha,x_\beta,y_\alpha,y_\beta,$ $u_\alpha,u_\beta,v_\alpha,v_\beta$. For these equations the initial value problem can be solved by the method of iterations** (at least in a neighborhood of the initial line Λ), and solving the characteristic system (7) is seen to

---

* For greater detail see Courant-Hilbert [12], Vol. II, Chapter V.

** See Courant-Hilbert [12], vol. II, Chapter V, Art. 7.

be equivalent to solving the initial value problem for our original system (1) (provided that the Jacobian $x_\alpha y_\beta - x_\beta y_\alpha$ does not vanish). Furthermore, this solution is uniquely determined.

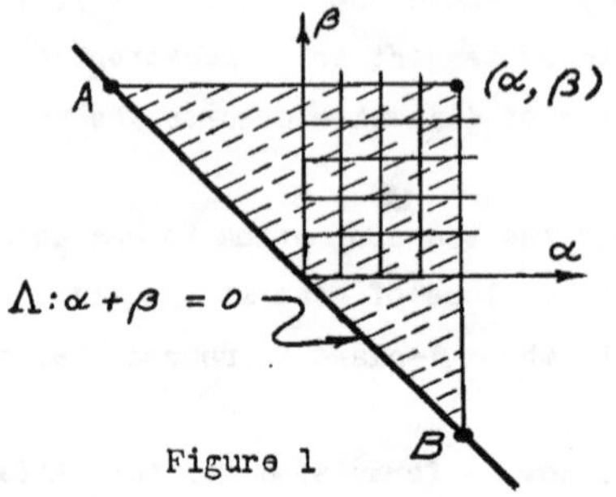

Figure 1

Most significant is the fact that the values u,v,x,y for the arguments α, β depend only on the initial values on Λ between the points A and B indicated on the diagram.

Since α = constant and β = constant are characteristics, our result has the following interpretation in the x,y-plane. The values of u,v at a point P in the x,y-plane do not depend on the totality of the initial values on L, but only on the initial values on the section of L intercepted by the two characteristics through P. This interval on the line L intercepted by the two characteristics is called the domain of dependence of the point P.

Correspondingly we may speak of the range of influence of a point Q on the line L, that is, the totality of points in the x,y-plane which are influenced by the initial data at the point Q. Evidently this is the angular region between the two characteristics drawn through Q. This range of influence of the point Q consists of all points P whose domains of dependence contain Q.

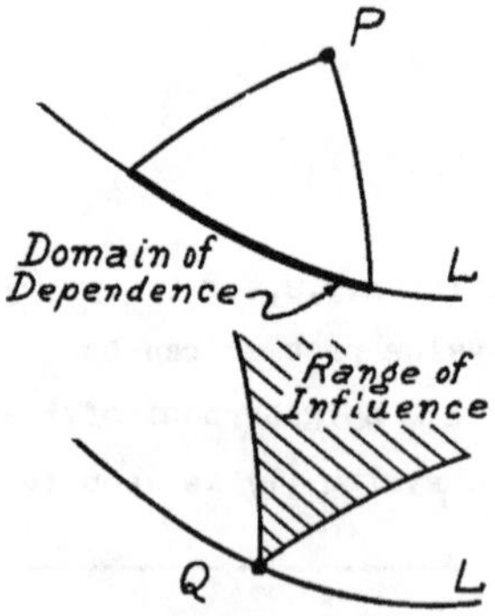

Figure 2

It is the existence of such domains of dependence and ranges of influence which characterizes phenomena of wave propagation in contrast to states of equilibrium. In the latter there is an organic connection between all the points of the medium. The differential equations are then elliptic and their solutions are analytic functions which are entirely determined by their values in any domain, however small. For problems of wave propagation, however, the solutions of the differential equations are not necessarily analytic. The rôle of the concepts of domain of dependence and range of influence is implicitly referred to in such expressions as "the medium at a point P does not know of the state at a point Q" (meaning that P does not belong to the range of influence of Q).

While the initial values of u and v along L are assumed continuous, discontinuities in the first or higher derivatives of these initial values may be permitted. At those points P whose domains of dependence do not contain the discontinuity points of the derivatives of the initial data, the solutions u,v have continuous first and higher derivatives.* From our preceding construction it could be inferred that such discontinuities of derivatives occur only along characteristics through the discontinuity points on the initial curve L (unless new discontinuities appear due to vanishing of the Jacobian $x_\alpha y_\beta - x_\beta y_\alpha$).

11. Propagation of discontinuities along characteristic lines. It is useful to enlarge upon the rôle of characteristics as possible loci of discontinuities. If at a point A there is a discontinuity in some derivatives of the initial data on L, then this discontinuity will be propagated along the two characteristics through A. It can never disappear.

In case the variable y is identified with the time t, this can be interpreted as follows. Any discontinuity will spread through the one-dimensional x-region with velocities $\frac{dx}{dt}$ given by the slopes of the two characteristics through the corresponding point of discontinuity in the x,t-plane.

In two-dimensional steady flow, characteristics issuing from the boundary of the flow will indicate small disturbances caused by slight roughness of the boundary which reach out into the medium. Such

---

* See Courant-Hilbert [12], vol. II, Chapter V, Art. 7.

characteristics, or Mach lines, are often actually visible in flows between slightly roughened walls.

The discontinuities spreading along characteristics can be described by the following mathematical considerations. Let us employ characteristic coordinates and suppose that the discontinuity appears across the line α = constant, so that the tangential derivatives with respect to β remain continuous. Then consider the two jumps

$$\left[u_\alpha\right]_{\alpha-0}^{\alpha+0} = U(\beta), \qquad \left[v_\alpha\right]_{\alpha-0}^{\alpha+0} = V(\beta),$$

where $u = u(\alpha,\beta)$.* We may assume that not only x, y, x_β, y_β but also x_α and y_α as well as $x_{\alpha\beta}$ and $y_{\alpha\beta}$ are continuous across the line α = constant. (This can be achieved by proper choice of the parameter α, replacing, if necessary, α by $\alpha' = W(\alpha)$ with a suitable W). We can now establish two homogeneous linear differential equations along α = constant for the discontinuity intensities $U(\beta)$ and $V(\beta)$.

Consider first equation II_+ on both sides of the characteristic α = constant, and subtract these equations from each other. Since the coefficients and derivatives with respect to β are continuous, we conclude that

$$U(\beta) + G_1(\beta)V(\beta) = 0, \tag{8}$$

where $G_1(\beta) = R\zeta_+ - S$ is a known function of β along α = constant. To obtain information from II_- we first differentiate with respect to α and carry out the previous process again. We then find a differential equation of the form

$$U_\beta + G_2V_\beta + MU + NV = 0, \tag{9}$$

where G_2, M, N are known functions of β along α = constant.**

* The use of characteristic coordinates will not be affected by the presence of discontinuities in u_x, u_y, v_x, v_y as long as u, v, x, y are continuous. Incidentally, the following considerations would not apply to discontinuities in u and v.

** Note that differentiation of G_2 with respect to α leads to $G_{2_u}[u_\alpha] + G_{2_v}[v_\alpha]$, etc.

The last two equations define each of the discontinuities U and V as a solution of a linear homogeneous ordinary differential equation. Hence these discontinuities are uniquely determined and are different from zero along the whole characteristic if they are known and different from zero at any point of the characteristic.

12. Characteristic lines as separation lines between regions of different types of flow. A remark of basic importance might be made here. Whenever the flow in two adjacent regions is described by expressions which are analytically different (e.g., when one is a region of rest or constant state while in the other region the state is not constant), then the two regions are necessarily separated by a characteristic. This statement is an immediate consequence of the fact that only along characteristics can derivatives of u and v of any order change discontinuously, or again, of the uniqueness theorem for the initial value problem for non-characteristic initial curves.*

If the differential equations are elliptic, no characteristics exist** and consequently no discontinuities of any type can occur. It is shown that in this case the solutions must be analytic functions of x and y and therefore cannot be constant in any region without being constant throughout.

13. Characteristic initial values. Along a curve L which is not characteristic for initial values of u,v, our differential equations, as we have seen, permit the calculation of the derivatives of u and v (and similarly of all higher derivatives), and determine the solution uniquely on both sides.

What corresponding information do the differential equations yield for a line L with values u,v which make it characteristic? The answer is found immediately from the normal form I,II of the equations.

---

* See Courant-Hilbert [12], vol. II, p. 297.

** Ibid., Chapter III, Art. 2 and p. 295.

Suppose L is a line C_+ or β = constant. Then II_+ shows that along L the values of u,v cannot both be prescribed arbitrarily but rather establishes a relation between them since it is an ordinary differential equation in u and v. We are consequently at liberty to prescribe only one function, e.g., u and, at a single point, the value of the other, v.

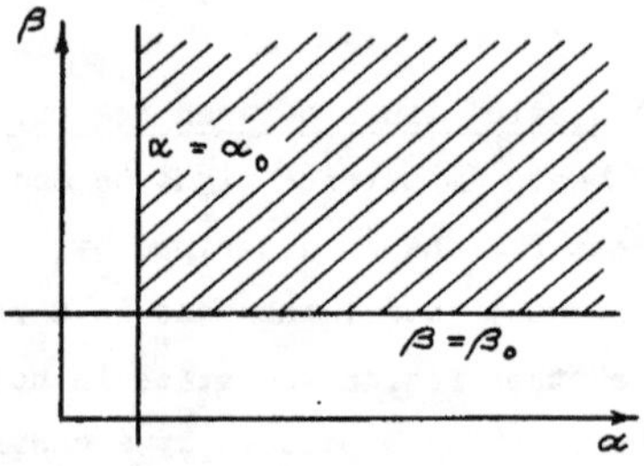

Figure 3

In many important applications the initial value problem is posed not for a non-characteristic initial curve L, but for initial data along two intersecting characteristic arcs. This characteristic initial value problem is formulated for the characteristic differential equations as follows. Given compatible values of u and v along two characteristic segments $\alpha = \alpha_0$, $\beta = \beta_0$ as indicated in Fig. 3, find the solution of I and II with these initial values for points α, β in one of the four angular domains, e.g., $\alpha > \alpha_0$, $\beta > \beta_0$. The solution is again uniquely determined and is obtained by the iteration method described above in Art. 10.

14. Relation between the characteristics in the x,y-plane and in the u,v-plane. The characteristics C and Γ can be brought into a simple geometric relation if we make the x,y-plane and the u,v-plane coincide along their respective coordinate axes and if we restrict ourselves to the special case where one of the equations (1) is $u_y - v_x = 0$ and where the other is homogeneous. This occurs when the flow is two-dimensional, isentropic, irrotational and steady. Then, as is easily verified, equations I and II (Art. 9) become

$$(11) \qquad \begin{cases} y_\alpha = \zeta_+ x_\alpha, & u_\alpha = -\zeta_- v_\alpha, \\ y_\beta = \zeta_- x_\beta, & u_\beta = -\zeta_+ v_\beta, \end{cases}$$

and it follows that

$$(12) \qquad \begin{cases} x_\alpha u_\beta + y_\alpha v_\beta = 0 \\ x_\beta u_\alpha + y_\beta v_\alpha = 0 . \end{cases}$$

This means that if x,y and u,v are represented in the same coordinate system, the directions of C_+ and Γ_- and of C_- and Γ_+ respectively through corresponding points (x,y) and (u,v) are perpendicular.

15. Application to the differential equations of gas dynamics. The preceding results form the basis for a more or less complete treatment of the equations of gas dynamics in the cases (a) - (d) characterized in Art.8. In all these cases the reduction of the differential equations to their characteristic form is immediately carried out, opening the way for theoretical as well as for numerical procedure. In Chapters III, IV and V we shall supplement the general theory for the different categories of flow problems under consideration. For the present we shall continue the general theory by discussing a point of major importance in applications, the notion of "simple wave".

16. Flow adjacent to a region of constant state. Simple waves. Very often the following situation is encountered. In a certain region (I) of the x,y-plane velocity, pressure and density are constant while this zone of constant state is followed by another zone (II) in which velocity, pressure and density vary. Then, as we saw, the two zones are necessarily separated by characteristics C or Mach lines. While restrictive conditions for a flow adjacent to a zone of constant state cannot be formulated in simple terms for every type of flow, there is an important class of flows in which the description of such flow patterns (II) is particularly simple. These are the cases where the differential equations (1) are homogeneous and have coefficients which do not contain x and y explicitly. Under these conditions, which are satisfied in cases (a) and (c), for one-dimensional flow and steady isentropic

irrotational two-dimensional flow, respectively, the characteristics Γ in the u,v-plane are two fixed families of curves which can be represented in the form $\beta(u,v)$ = constant and $\alpha(u,v)$ = constant. In the following pages we shall make the assumption of fixed characteristics in this sense. (Always, we assume the hyperbolic character of the differential equations, which in the case (c) is tantamount to the supersonic character of the flow.) Then the flow (II) is of peculiar pattern, called a simple wave, which we shall presently define.

For any flow, the characteristics C_+ and C_- in a zone of constant state are straight lines, since constant values of u and v imply constant values of β and α. For flows with fixed characteristics we define a simple wave as a zone with the following property: One of the two sets of characteristics C, say the family of curves C_+, consists of straight lines along each of which the values of u, v, p, ρ remain constant, while these constants vary from one characteristic C to another.

The situation thus described can be interpreted by reference to the hodograph plane with the coordinates u,v. In general, i.e., when $\Delta = u_x v_y - u_y v_x \neq 0$, there corresponds to a sufficiently small domain of the flow in the x,y-plane an image domain in the u,v-plane, and we may introduce u and v as independent variables (and thus linearize our differential equations, as seen in Art. 8). For a domain of constant state in the x,y-plane, however, the image in the hodograph plane is a single point, which makes the interchanging of dependent and independent variables impossible and of course implies $\Delta = 0$. Likewise, a domain (II) of a simple wave is not mapped on a domain of the u,v-plane as in the general case, but only on a one-parametric set of points (u,v), each corresponding to a whole C_+ characteristic of the simple wave in the x,y-plane of the flow. Thus, for simple waves we again have $\Delta = 0$, and linearization is impossible. As a remarkable fact we shall find that the point sets (u,v) corresponding to a simple wave are restricted to an arc of a Γ characteristic in the hodograph plane.

We shall now supply the mathematical proof and some additional facts. This is almost immediately achieved on the basis of the characteristic forms I and II of our differential equations (1), which we may write

$$I_+: \; y_\alpha = \xi_+ x_\alpha \qquad I_-: \; y_\beta = \xi_- x_\beta$$

$$II_+: \; u_\alpha = \eta_+ v_\alpha \qquad II_-: \; u_\beta = \eta_- v_\beta$$

where ξ_+, ξ_-, η_+, η_- are known functions of u and v. Since $\eta_- - \eta_+ = R(\xi_+ - \xi_-)$ with $R \neq 0$, the relation $\eta_+ - \eta_- \neq 0$ follows from $\xi_+ - \xi_- \neq 0$, the latter being a consequence of the hyperbolic character of our equations.

Now let us assume that $\Delta = u_\alpha v_\beta - u_\beta v_\alpha = 0$ in a region of the x,y-plane. Then from II we conclude that

$$(\eta_+ - \eta_-) v_\alpha v_\beta = 0$$

and hence $v_\alpha v_\beta = 0$. Suppose that the factor v_α vanishes so that $v = v(\beta)$ is a function of only one of the characteristic parameters. Then from II_+ we also infer that $u_\alpha = 0$ so that $u = u(\beta)$ likewise depends on the same characteristic parameter β alone, which means that u and v are constant along C_+. Unless u and v are also independent of β so that the state is constant, the one-parametric set of points (u,v) will, according to II_-, satisfy the ordinary differential equation $\frac{du}{dv} = \eta_-$ and will constitute an arc of a characteristic Γ_- in the u,v-plane.

We now make the general observation that if the values of u and v are constant on a characteristic C_+, then this C_+ is a straight line. For on C_+ the value ξ_+ will be constant, ξ_+ depending only on u and v according to our assumption; and therefore $\frac{dy}{dx} = \xi_+$ defines a fixed slope for the curve C_+.

As a consequence, we infer that our region of the x,y-plane is covered by a family of straight characteristics C_+ depending on the parameter β, each carrying a fixed set of values u,v. If the point (u,v) is the same for the whole region of the flow plane, then this region is one of constant state. Otherwise the values u and v depend on the parameter β, and the point $u(\beta)$, $v(\beta)$ describes an arc of a Γ_- characteristic in the hodograph plane, so that our region is then covered by a genuine simple wave.

There exists a great variety of simple waves satisfying our differential equations, and it is of interest to ascertain what data are suitable for specifying an individual simple wave. One mathematical possibility is to prescribe arbitrarily the family of straight characteristics C_+ and in addition to select arbitrarily the Γ_- characteristic on which the (u,v) values of the wave lie.* From the physical point of view, however, it is more natural to prescribe other data, such as one streamline and the flow speed at one point of it.

In the special case where all the straight characteristics C_+ diverge from a point O, we call the wave <u>centered</u> with the center O. Such a <u>centered wave</u>, as we shall see, is determined if we know the velocity on one of the lines C_+.

The role of simple waves as zones necessarily neighboring zones of constant state (in a flow with fixed characteristics Γ) is established by the following <u>fundamental theorem</u>: If on a characteristic line C_+^0 of the differential equations (1) the values u_0, v_0 remain constant, then C_+^0 is a straight line and is embedded in a family of straight line characteristics C_+ on each of which u and v remain constant. (Hence C_+^0 is either embedded in a zone of constant state or in a simple wave, or forms the boundary between two such zones). This may be proved as follows. First, C_+^0 is a straight line by I , since ξ_+ is constant on C_+^0. Now let C_+ be another characteristic sufficiently close to C_+^0 so that the characteristics C_- through any two points A and B on C_+ will intersect C_+^0 in two points A_0 and B_0. Now we make use of the fact

---

* To construct the wave, i.e., to find the corresponding solution u(x,y), v(x,y) of (1), one proceeds as follows. To each β characterizing $u(\beta)$, $v(\beta)$ on Γ_-, we determine $\xi_+[u(\beta),v(\beta)]$. Then we select from the given family of lines C_+ in the x,y-plane those having the slope ξ_+ and attach to these lines the values $u = u(\beta)$, $v = v(\beta)$. Thus u(x,y) and v(x,y) are constructed. That we really have obtained a solution is easily verified.

that the characteristics $\Gamma_{\pm}$ are fixed and can be represented in the forms $\beta(u,v)$ = const. and $\alpha(u,v)$ = const., respectively. Since

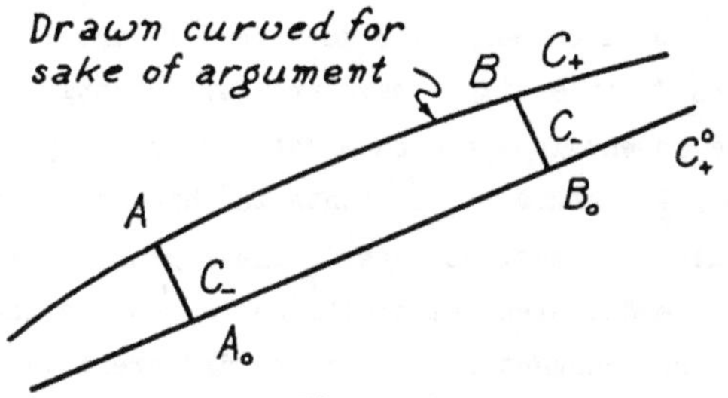

Figure 4

the images of the points A and A_0 in the u,v-plane lie on the same characteristic Γ_- and since the same is true for the points B and B_0, while the images of the points A and B are on the same characteristic Γ_+ , we have, in an obvious notation

$$\alpha(A) = \alpha(A_0), \qquad \alpha(B) = \alpha(B_0), \qquad \beta(A) = \beta(B)$$

Furthermore, according to our assumptions, the point in the u,v-plane corresponding to A_0 is the same as that corresponding to B_0. Hence we have

$$\alpha(A_0) = \alpha(B_0)$$

the value of both of the functions α being determined by the two arguments u,v. Consequently we have

$$\alpha(A) = \alpha(B),$$

which, together with

$$\beta(A) = \beta(B),$$

implies that

$$u(A) = u(B), \quad v(A) = v(B).$$

In other words, u and v are constant on C_+; and, as a consequence, C_+ must also be straight, which proves our theorem.

In the subsequent analysis of various types of flows we shall discuss simple waves in greater detail. Two remarks might be added here, however.

The first remark concerns the fact that if the lines of a system of characteristics C_+ in a simple wave are sufficiently extended, in general they have an envelope (which might possibly degenerate into a point). Obviously the geometrical points in which two different characteristics C_+ intersect must be outside the region for which the simple wave solution of the differential equations represents the actual flow, for the two different characteristics carry different values of u and v, and thus at a point of intersection a physical absurdity would arise. Therefore the actual simple wave can never reach beyond the envelope in the x,y-plane. The solution can be geometrically interpreted by considering the mapping of the α,β-plane onto the x,y-plane by means of a simple wave solution $x = x(\alpha,\beta)$, $y = y(\alpha,\beta)$. As long as the mapping is one-to-one the simple wave solution has a physical meaning; but this is no longer true when in the mapping of the α,β-plane into the x,y-image plane a part of the α,β-plane is folded over, so that a sector of the image plane will be covered three times. In this case the edges of the fold then form the envelope of the lines C_+ and it is clear that the solution ceases to represent a definite physical state.

The second remark concerns the fact that constant states as well as simple waves are merely convenient idealizations of actual phenomena, and that of the various types of flows under consideration only the general flow in one dimension and the steady isentropic flow in two dimensions admit of simple waves as studied in this chapter. In all other cases it is no longer true that the flow in a region adjacent to a zone of constant state is covered by characteristics along which the velocity and density are constant. The fundamental problem of determining the flow in such a region must be answered by a direct attack on the corresponding characteristic initial value problem of the differential equations.

III. MOTION IN ONE DIMENSION

A. Integration of the Differential Equations

17. Introductory remarks. Motion of compressible fluids admits of a fairly exhaustive mathematical treatment if the state of the medium depends only on the time t and on a single space coordinate x. In this case we speak of motion in one dimension or one-dimensional motion. The differential equations of motion then reduce to the simple systems (A), (B), (C) or (D) of Chapter II, Art. 8, and a complete integration in terms of arbitrary functions becomes possible.

As the model of one-dimensional motion we shall usually consider the flow of a gas in a long tube extending along the x-axis.* The tube may be infinite, semi-infinite or finite, i.e., open at both ends, closed by a piston at one end, or by pistons or walls at both ends. Unless otherwise stated, we shall assume an initial state of uniform velocity u_0, say rest, $u_0 = 0$, and uniform pressure p_0 and density ρ_0. The motion will then be caused by the action of the pistons at the ends.

It is convenient to represent the phenomena in an x,t-coordinate system, and to refer to "paths of particles" in the x,t-plane. The x-coordinate of the piston at the left end of the gas-filled tube may be $x = 0$ for $t = 0$. Then the motion of the piston is represented in the x,t-plane as a curve L, the "piston curve", starting at the origin as indicated in Fig. 1(a) for compressive, or as in Fig. 1(b) for expansive action of the piston.

In Section B of the present chapter we shall study the simplest types of continuous motion of a gas, in particular, the so-called rarefaction waves or expansion waves, caused by receding pistons. Section C is devoted to a discussion of motion involving shock discontinuities, which

---

* This problem was treated by Hugoniot and later by Rayleigh. See Rayleigh [16] in the Bibliography.

may develop as a result of <u>compression</u>. With what may be some sacrifice of conciseness, an attempt is made to illuminate the shock wave theory

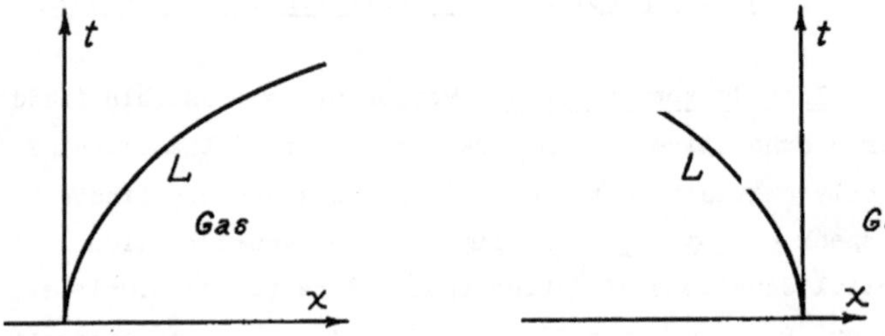

(a) Piston curve. Compressive action.

(b) Piston curve. Expansive action.

Figure 1

from various angles. In Section D it is shown how more general types of motion result from interaction of the elementary motions studied in Sections B and C.

18. <u>Solution of the differential equations</u>. The differential equations of one-dimensional motion (see Chapter II, Art. 8) can be integrated almost immediately when written in their characteristic form. Replacing the quantity y by t and the quantity v by ρ in the theory of Chapter II, Art. 9, we find, with $c^2 = p'(\rho)$, the expressions $\zeta_+ = \frac{1}{u + c}$ and $\zeta_- = \frac{1}{u - c}$ (by a straight-forward application of the procedure described there); accordingly, the characteristic form of the differential equations with the characteristic parameters α, β as independent variables is

$$(1) \quad \begin{cases} I_+: \quad x_\alpha = (u + c)t_\alpha \qquad II_+: \quad u_\alpha = -\frac{c}{\rho}\rho_\alpha \\ I_-: \quad x_\beta = (u - c)t_\beta \qquad II_-: \quad u_\beta = \frac{c}{\rho}\rho_\beta \end{cases}$$

The system II can be completely solved for u, ρ by the relations

$$(2)\qquad \begin{cases} u + G(\rho) = V(\beta) = 2r \\ u - G(\rho) = W(\alpha) = 2s \end{cases} ,$$

where V and W (and r and s) are arbitrary functions and where $G(\rho)$ is defined by

$$(3)\qquad G(\rho) = \int_{\rho'}^{\rho} \frac{c}{\rho} d\rho ,$$

ρ' being an arbitrary constant. Instead of α and β it is often convenient to introduce the special parameters r and s so that

$$(4)\qquad u = r + s, \qquad G(\rho) = r - s;$$

then u, ρ and c are known functions of r and s.

Substituting these solutions of II in I, we obtain for x and t as functions of α and β, or of r and s, two linear differential equations of first order.

The structure of the solutions is best described by reference to the characteristics β = constant, α = constant (or r = constant, s = constant), which in the x,t-plane are characteristics C_+ and C_- respectively, and in the u,ρ-plane characteristics Γ_+ and Γ_- respectively. Then we have

$$(5)\qquad \begin{cases} \begin{matrix}\beta = \text{constant} \\ \text{or} \\ r = \text{constant}\end{matrix} & \begin{cases} \text{On } \Gamma_+: & u + G(\rho) = 2r \\ \text{On } C_+: & \dfrac{dx}{dt} = u + c \end{cases} \\ \\ \begin{matrix}\alpha = \text{constant} \\ \text{or} \\ s = \text{constant}\end{matrix} & \begin{cases} \text{On } \Gamma_-: & u - G(\rho) = 2s \\ \text{On } C_-: & \dfrac{dx}{dt} = u - c \end{cases} \end{cases}$$

Thus for polytropic gases with $\rho' = 0$ we have $r = \frac{u}{2} + \frac{c}{\gamma - 1}$, $s = \frac{u}{2} - \frac{c}{\gamma - 1}$ and we infer the following basic statement:

$$(6)\quad \begin{cases} \text{On } C_+\text{: } \dfrac{dx}{dt} = u + c, \text{ and the value of } \dfrac{u}{2} + \dfrac{c}{\gamma - 1} \text{ remains constant.} \\ \text{On } C_-\text{: } \dfrac{dx}{dt} = u - c, \text{ and the value of } \dfrac{u}{2} - \dfrac{c}{\gamma - 1} \text{ remains constant.} \end{cases}$$

If we consider not u and ρ, but u and the sound speed c as dependent variables, the characteristics Γ become straight lines:*

$$(7)\quad \begin{cases} \Gamma_+\text{: } \dfrac{u}{2} + \dfrac{c}{\gamma - 1} = \text{constant} \\ \Gamma_-\text{: } \dfrac{u}{2} - \dfrac{c}{\gamma - 1} = \text{constant} \end{cases}$$

with $c \geq 0$.

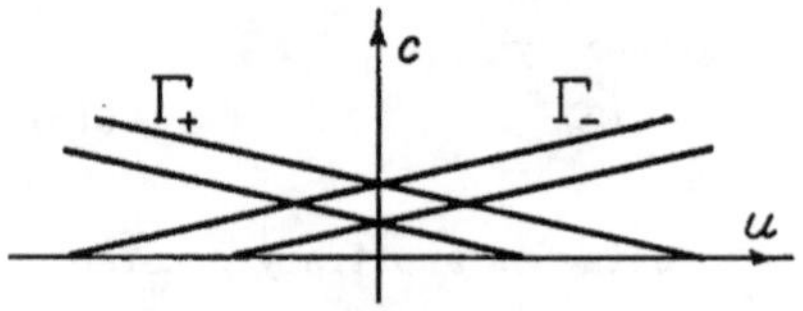

Figure 2
Characteristics
in u,c-plane.

---

* The characteristics Γ_+ and Γ_- are fixed curves in the u,ρ-plane, namely,

$$(8)\quad \begin{cases} \Gamma_+\text{: } u + \dfrac{2\sqrt{A\gamma}}{\gamma - 1}\rho^{\frac{\gamma-1}{2}} = \text{constant} \\ \Gamma_-\text{: } u - \dfrac{2\sqrt{A\gamma}}{\gamma - 1}\rho^{\frac{\gamma-1}{2}} = \text{constant} \end{cases}$$

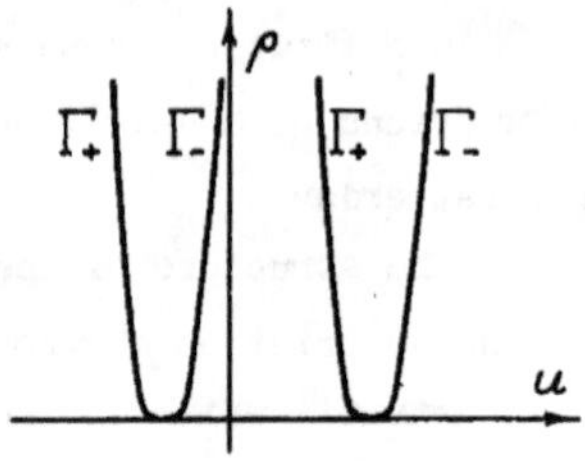

Figure 3
Characteristics
in u,ρ-plane.

as shown in Fig. 3, where the left-hand branches represent the curves Γ_+ and the right-hand branches the curves Γ_-.

With u and the enthalpy

$$i = \int_0^\rho \frac{c^2}{\rho}\, d\rho = \frac{c^2}{\gamma - 1}$$

as dependent variables, the characteristics Γ become ordinary parabolas in the u,i-plane:

$$(9)\quad \begin{cases} \Gamma_+\text{: } u = -2\sqrt{\dfrac{i}{\gamma - 1}} + \text{constant} \\ \Gamma_-\text{: } u = 2\sqrt{\dfrac{i}{\gamma - 1}} + \text{constant,} \end{cases}$$

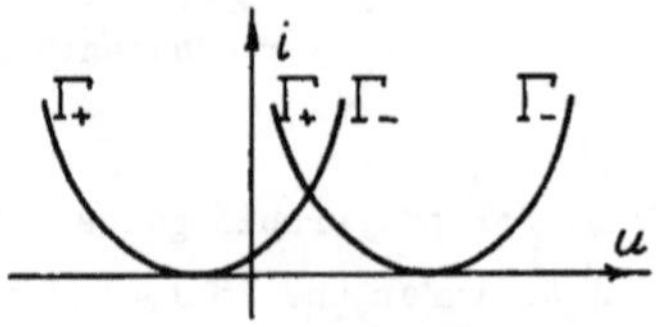

Figure 4
Characteristics
in u,i-plane.

19. Simple waves. Our main subject in the present and next few articles will be the motion caused by a piston moving in a gas which is initially at rest.

No matter whether the piston recedes from or moves into the gas, not all parts of the gas will be affected instantaneously. There will be a "wave" proceeding from the piston into the gas and only the particles which have been reached by the wave front will be disturbed from their initial state of rest. If this wave represents a continuous motion, as is always the case if the piston recedes from the gas, the wave front progresses with the sound speed c_0 of the quiet gas. If the piston moves into the gas the situation may become more complicated through the emergence of a supersonic discontinuous shock wave as we shall see in Section C. At any rate, in Sections A and B we confine ourselves to a consideration of continuous motions satisfying the differential equations (at least near the piston). Such a continuous motion of the gas can be completely determined by the simple wave theory of Chapter II, Art. 16.

---

where the left-hand branches of the parabolas represent Γ_+, the right-hand branches Γ_-. For $i \leq 0$, naturally, as for $\rho \leq 0$ or $c \leq 0$, the differential equations lose their physical meaning.

No essentially new elements occur in the Lagrangean representation (see Chapter I, Art. 7). With the independent variables $h = \int_{x_0}^{x} \rho(\xi)d\xi$ (instead of x) and t (instead of y), with the dependent variables u and τ (instead of v), and with $k(\tau) = \rho c = \frac{c(\tau)}{\tau}$, the equations are

$$(10) \qquad \text{I}\begin{cases} C_+: & h_\alpha = k(\tau)t_\alpha \\ C_-: & h_\beta = -k(\tau)t_\beta \end{cases} \qquad \text{II}\begin{cases} \Gamma_+: & u_\alpha = k(\tau)\tau_\alpha \\ \Gamma_-: & u_\beta = -k(\tau)\tau_\beta \end{cases}$$

The characteristics Γ_+ and Γ_- again can be explicitly described:

$$(11) \qquad u \mp \int_0^\tau k(\tau)d\tau = \text{constant}.$$

In the x,t-representation the undisturbed gas corresponds to a zone of rest (I) adjacent to the x-axis and (as long as the disturbance proceeds at sound speed) bounded by a characteristic C_+: $x = c_0 t$, which terminates the range of influence of the piston curve L. According to the general theory of Art. 16, the zone (I) of rest is followed beyond the line $x = c_0 t$ by a simple wave (II) generated by a family of straight characteristics C_+.

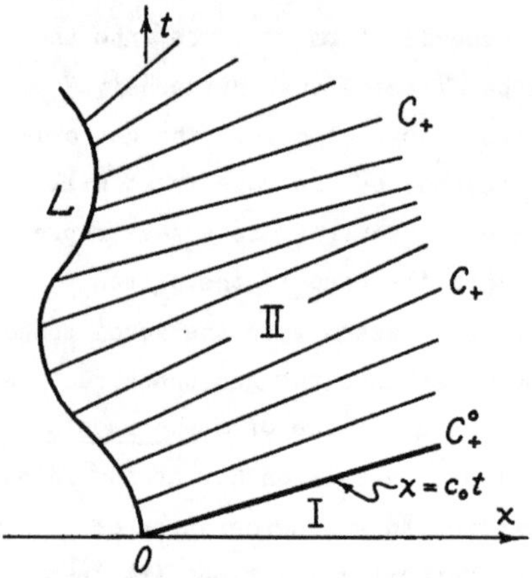

Figure 5
Simple wave (II) adjacent to zone of constant state (I).

We shall here give a brief description of the simple waves and then in Sections B and C supply the details of the two cases of receding and impinging piston motions separately.

Along each conjugate characteristic C_-, which cuts across the lines C_+, we have, according to equation (5), Art. 18, the relation

$$u - G(\rho) = 2s = \text{constant}$$

where the constant is determined by the consideration that all the characteristics intersect an "initial characteristic" C_+^0 with values u_0, ρ_0 (e.g., the characteristic terminating the zone of constant state of rest (I)). Thus, throughout the simple wave we have

$$u - G(\rho) = u_0 - G(\rho_0) \tag{12}$$

and, in particular, if the initial characteristic terminates a state of rest,

$$u - G(\rho) = -G(\rho_0). \tag{13}$$

For polytropic gases we have $G(\rho) = \frac{2c(\rho)}{\gamma - 1}$, therefore the basic relation in a forward-facing simple wave (see (7), Art. 18) is

(14) $$u - \frac{2}{\gamma - 1} c = u_o - \frac{2}{\gamma - 1} c_o$$

or, in particular, if the initial state (o) is a state of rest,

(15) $$u - \frac{2}{\gamma - 1} c = -\frac{2}{\gamma - 1} c_o$$

where c_o is the sound speed in the quiet gas. With the abbreviation

(16) $$\mu^2 = \frac{\gamma - 1}{\gamma + 1}, \qquad 1 - \mu^2 = \frac{2}{\gamma + 1},$$

our relation can be written in the form

(17) $$\mu^2(u - u_o) = (1 - \mu^2)(c - c_o).$$

Incidentally, these last equations are nothing but the equation of the single characteristic Γ_- in the u,c-plane which belongs to the simple wave in accordance with the general theory of Chapter II, Art. 16, and which happens to be a straight line also.

B. Continuous Motion. Rarefaction Waves.

20. Rarefaction waves. We now distinguish between the cases of expansive and compressive motion and consider first the case of expansive motion caused by a receding piston, assuming from the outset that the medium is a polytropic gas originally at rest with constant density ρ_0 and sound speed c_0. Furthermore, it is assumed that the piston, originally at rest, is withdrawn with increasing speed until ultimately the constant velocity -w is attained. Then the piston curve L will bend backward from 0 to a point B where the slope -w with respect to

the t-axis is reached and then continue as a straight line in the same direction, as shown in Fig. 6. We then have a zone (I) of rest: $0 \leq c_0 t \leq x$, and an adjacent zone (II) of a simple wave, which we shall call an <u>expansion wave</u> or <u>rarefaction wave</u> because, as we shall see, the gas flowing through the zone of this wave steadily decreases in density, and even at a fixed point of the tube the density decreases as long as the point remains in the zone of the simple wave. The particles flow toward the receding piston, starting with zero velocity at the <u>head</u> or <u>front</u> of the rarefaction wave, which is represented by the characteristic C_+^0: $x = c_0 t$, and proceed into the zone of constant state with the speed of sound. In this wave the straight lines of the generating family of characteristics C_+ start at the piston curve L: $x = f(t)$.

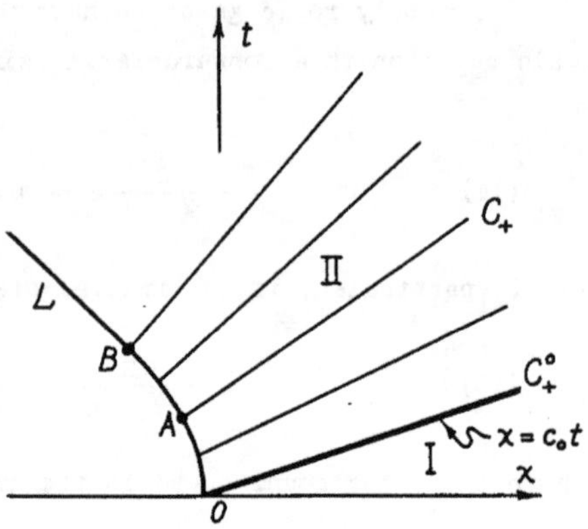

Figure 6
Rarefaction wave (II) resulting from expansive action of piston.

To construct the wave and to prove and amplify the preceding statement we consider a point A on the piston curve L from which a characteristic C_+ is assumed to start into zone (II). At A the velocity u_A of the gas is known, since it is equal to the velocity $\frac{dx}{dt} = f'(t)$ of the piston. We also obtain the value of the sound speed c_A at A by formula (14) of Art. 19:

$$u_A - \frac{2}{\gamma - 1} c_A = u_0 - \frac{2}{\gamma - 1} c_o \qquad (18)$$

Now the straight line C_+ through A is determined by its slope $\frac{dx}{dt} = u_A + c_A$; and on C_+ the values of u, c, ρ, p are now fixed as u_A, c_A, ρ_A, p_A where ρ_A, p_A are given by

(19) $$\frac{p_A}{p_0} = \left(\frac{\rho_A}{\rho_0}\right)^{\gamma}, \qquad c_A^2 = \gamma\frac{p_A}{\rho_A},$$

which leads immediately to the important relations

(20) $$p = p_0\left[1 + \frac{\gamma - 1}{2}\frac{u - u_0}{c_0}\right]^{\frac{2\gamma}{\gamma - 1}},$$

(21) $$\rho = \rho_0\left[1 + \frac{\gamma - 1}{2}\frac{u - u_0}{c_0}\right]^{\frac{2}{\gamma - 1}},$$

where the subscript A is omitted. With reference to the state on any initial characteristic C_+^0, these express density and pressure in the simple wave in terms of the velocity. If we choose as this initial characteristic the one terminating the state (I) of rest, we have $u_0 = 0$ and obtain

(22) $$p = p_0\left[1 + \frac{\gamma - 1}{2}\frac{u}{c_0}\right]^{\frac{2\gamma}{\gamma - 1}},$$

(23) $$\rho = \rho_0\left[1 + \frac{\gamma - 1}{2}\frac{u}{c_0}\right]^{\frac{2}{\gamma - 1}},$$

(24) $$c = c_0 + \frac{\gamma - 1}{2}u.$$

Since u is negative in the simple wave, these formulas exhibit the fact that density and pressure decrease as we follow the path of a particle represented in the x,t-plane by a trajectory of the family of characteristics C_+.* For increasing t a fixed point in the gas will

---

* For air $\gamma = 1.4$ and the exponents are 7 and 5 respectively, which shows that the decrease in pressure and density in an expansion wave is rather rapid.

belong to "later" characteristics C_+, i.e., larger values of $-u$, and hence, as stated before, to smaller values of p and ρ.

The law of rarefaction expressed by (22) and (23) becomes meaningless as soon as $-u > q_e$ where

(25) $$q_e = \frac{2}{\gamma - 1} c_o$$

is the escape speed of the gas originally at rest. If $-u$ reaches the escape speed, the rarefaction has thinned the gas down to zero density and pressure and the sound speed has likewise decreased to zero. If a rarefaction wave extends to this stage it is called a complete rarefaction wave as it then ends in a vacuum.

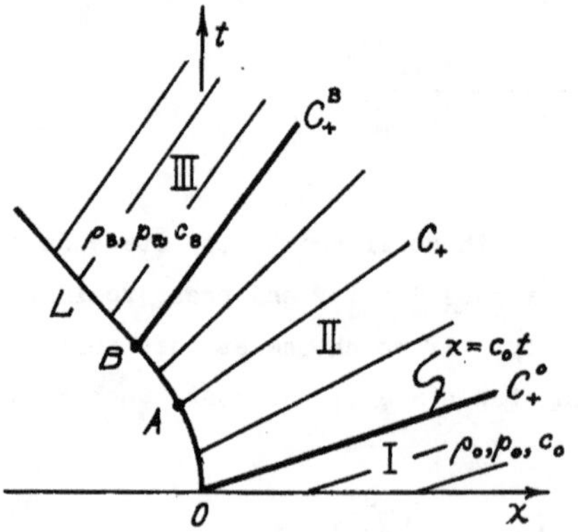

Figure 7
Simple wave (II) connecting two regions (I) and (III) of constant state ($w < q_e$).

For the end or tail of the expansion wave there result two possibilities, according to whether or not the terminal speed w of the piston is below the escape speed q_e.

If $w < q_e$, the preceding construction of the simple wave will yield characteristics C_+ for every point A on the piston curve from 0 to B. The rarefaction wave is incomplete and ends at the characteristic C_+^B through B with $u = u_B = -w$ and

$$c_B = c_o - \frac{\gamma - 1}{2} w,$$

(26) $$p_B = p_o \left(1 - \frac{w}{q_e}\right)^{\frac{2\gamma}{\gamma - 1}}$$

$$\rho_B = \rho_o \left(1 - \frac{w}{q_e}\right)^{\frac{2}{\gamma - 1}}$$

It is followed by a zone (III) of constant state u_B, ρ_B, p_B, c_B between the tail of the incomplete rarefaction wave and the piston in which the characteristics C_+ are all parallel (as they are in the zone (I) of constant state in front of the simple wave).

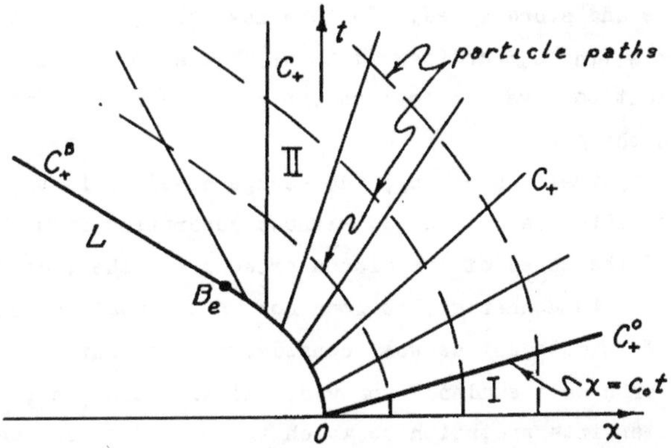

Figure 8
Complete rarefaction ($w = q_e$).

If $w = q_e$ the characteristic C_+^B through $B = B_e$ is tangent to the piston curve, for at B the piston curve has the slope $f'(t) = -w$, while that of the characteristic C_+^B is $\frac{dx}{dt} = u_B + c_B = -w = -q_e$ since $c_B = 0$. In other words, the wave is just completed at the piston.

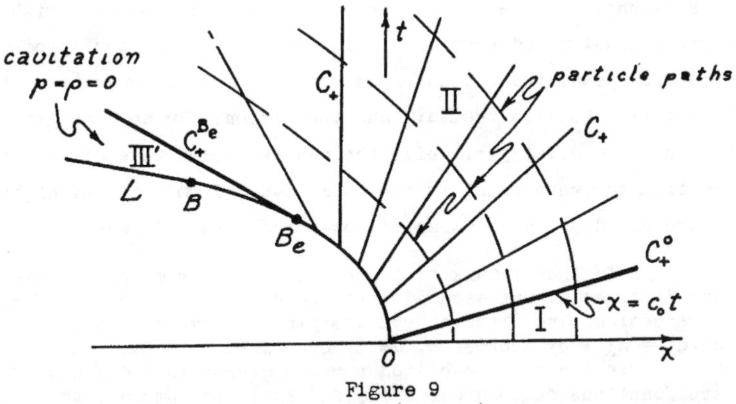

Figure 9
Cavitation ($w > q_e$).

If $w > q_e$ the completion of the wave is already achieved before the piston reaches the terminal speed. There will be a point B_e on the piston curve L between O and B for which the characteristic $C_+^{B_e}$ is tangent to the piston curve and carries the value zero for density, pressure and sound speed. In this case the rarefaction is completed with this line $C_+^{B_e}$ and beyond it we have a zone (III') of cavitation, equivalent to a vacuum between the receding piston and the tail of the wave in the gas.

Physically speaking, the escape speed q_e is the speed beyond which a piston cannot recede without separating from the thinned-out gas. If the speed of the piston exceeds q_e, then, as far as the motion of the gas is concerned, it does not matter what the actual value of w is. We might just as well consider w as infinite or imagine the piston or a wall suddenly removed, allowing the gas to escape into a vacuum, an interpretation to which the name "escape speed" alludes.

We can summarize our results qualitatively as follows. A piston receding from a gas at rest with speed which never decreases causes an expansion wave of particles moving toward the piston. At the head or front of the wave, which moves into the gas at sound speed, the velocity of the gas is zero. Through the wave the gas is accelerated. If the piston speed w is below the escape speed q_e, the gas will expand until it has reached the speed w of the piston and then continue with constant velocity, density and pressure. If, however, the piston speed exceeds the escape speed, the expansion is complete and the wave ends in a zone of cavitation between the tail and the piston. In any case the wave moves into the quiet gas, while the gas particles move at increasing speed from the wave front to the tail, i.e., from zones of higher pressure and density to zones of lower pressure and density.

A further remark of a general character might be added. In our diagrams it was assumed that the characteristics C_+ of the rarefaction wave diverge from the piston curve L, i.e., that $dx/dt = u_A + c_A$ decreases, as A moves from O to B.

Since along the piston curve ρ decreases and since u and c are functions of ρ in our wave, our statement amounts to

$du/d\rho + dc/d\rho > 0$. At the piston we have, from (12), $\frac{du}{d\rho} = \frac{c}{\rho}$, hence the preceding relation is equivalent to $\frac{1}{\rho} + \frac{1}{c}\frac{dc}{d\rho} > 0$ or $\frac{d}{d\rho}(\log \rho c) > 0$. In other words, a sufficient condition for the desired divergence of the lines C_+ (admitting a general equation of state) is that the <u>acoustic impedance</u> ρc increases with ρ, a condition certainly satisfied for polytropic gases.

Finally, it should be stated that in our simple waves the conjugate characteristics C_- and the paths of the gas particles can be found by integrating the ordinary differential equations

$$\frac{dx}{dt} = u - c \quad \text{and} \quad \frac{dx}{dt} = u,$$

respectively, after the functions $u(x,t)$ and $c(x,t)$ have been found by the previous construction.

In a complete rarefaction wave the characteristics C_- as well as the particle paths acquire asymptotically the direction of the last characteristic C_+^e, i.e., the direction $dx/dt = -q_e$.

21. <u>Centered rarefaction waves</u>. Of particular interest is the case of a <u>centered rarefaction wave</u>, which corresponds to an idealized piston motion where the acceleration from rest to a constant terminal velocity $-w$ takes place in an infinitely small time interval, i.e., instantaneously. Then the family of characteristics C_+ forming the simple wave will degenerate into a pencil of lines through the origin O: $x = 0$, $t = 0$ (see Fig. 10).

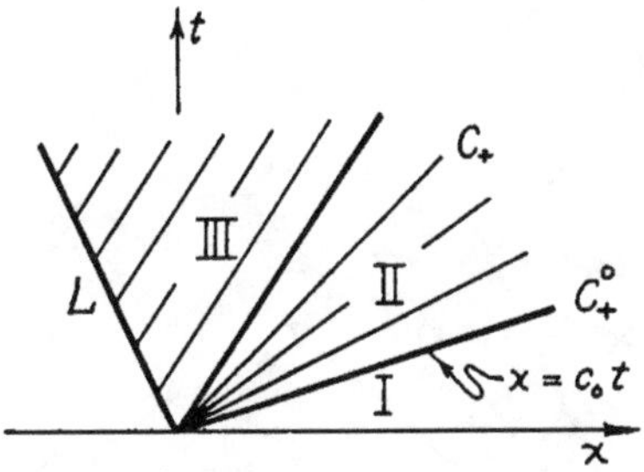

Figure 10
Centered rarefaction wave ($w < q_e$).

In the center O the quantities u, ρ, p as functions of x and t are discontinuous, but this <u>discontinuity is immediately smoothed out</u> in the subsequent motion. Here we have the first and typical example of an initial discontinuity which immediately resolves into continuous motion.

22. Explicit formulas for centered rarefaction waves. For centered rarefaction waves all quantities can be expressed explicitly as simple functions of x and t. With the center at the origin 0, we have for each line C_+ through 0 with flow velocity u

$$\frac{dx}{dt} = \frac{x}{t}\,.$$

Hence, by virtue of $\frac{dx}{dt} = u + c = u + (c_0 + \frac{\gamma - 1}{2} u)$ (see equations (6) and (24)), we have in the wave zone (II), $\frac{x}{t} = u + (c_0 + \frac{\gamma - 1}{2} u)$ or

$$(27) \qquad u = \frac{2}{\gamma + 1}\left(\frac{x}{t} - c_0\right) = (1 - \mu^2)\left(\frac{x}{t} - c_0\right)$$

where $\mu^2 = \frac{\gamma - 1}{\gamma + 1}$; and from $c = c_0 + \frac{\gamma - 1}{2} u$ we obtain

$$(28) \qquad c = \mu^2 \frac{x}{t} + (1 - \mu^2)c_0 .$$

Thus u and c are known explicitly in the zone (II) of the rarefaction wave, and p and ρ can be found (see (22), (23), (25)) by using

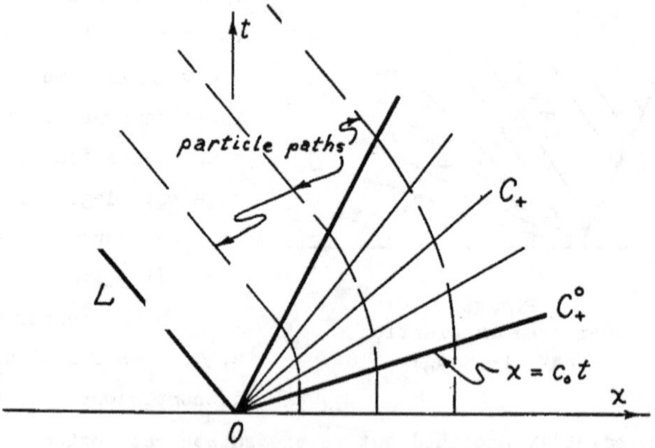

Figure 11
Particle paths in flow involving centered rarefaction wave ($w < q_e$).

$$(29) \qquad p = p_0\left(1 + \frac{u}{q_e}\right)^{\frac{2\gamma}{\gamma - 1}}, \qquad \rho = \rho_0\left(1 + \frac{u}{q_e}\right)^{\frac{2}{\gamma - 1}}$$

We can find the particle paths in (II) by integrating $\frac{dx}{dt} = u = (1 - \mu^2)\left(\frac{x}{t} - c_0\right)$ (see equation (27)). Upon making the substitution $x = \xi(t) - \frac{2c_0}{\gamma - 1}t = \xi(t) - \frac{1 - \mu^2}{\mu^2}c_0 t$, we obtain $\frac{d\xi}{dt} = (1 - \mu^2)\frac{\xi}{t}$, and by integrating we find that $\xi = At^{1-\mu^2}$, where A is an arbitrary constant, or

$$(30) \qquad x = t\left\{At^{-\mu^2} - \frac{2c_0}{\gamma - 1}\right\}$$

This formula is valid as long as the particles remain in the rarefaction zone (II).

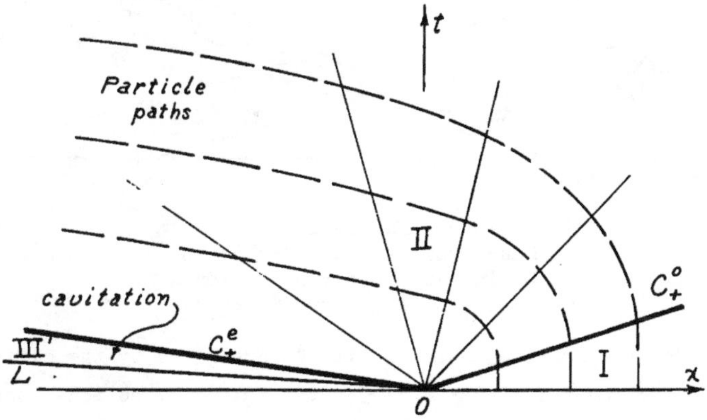

Figure 12
Cavitation behind centered rarefaction wave ($w > q_e$).

In the case of a <u>complete rarefaction</u> ending with zero density for $w \geq q_e$, formula (30) holds for arbitrarily large values of t, and we have, for large t, the asymptotic expression

$$(31) \qquad x \sim -\frac{2c_0}{\gamma - 1}t$$

As remarked previously, the fluid remains in the zone (II) of rarefaction, and in the x,t-diagram the particle paths acquire asymptotically the direction of the characteristic C_+^e on which the escape speed q_e is attained (see Fig. 12).

For $w < q_e$ the rarefaction wave terminates at the characteristic C_+ on which the velocity has the value $u = -w$, and all the particle paths emerge from (II) parallel to the terminal direction of the piston curve L and remain parallel in zone (III).

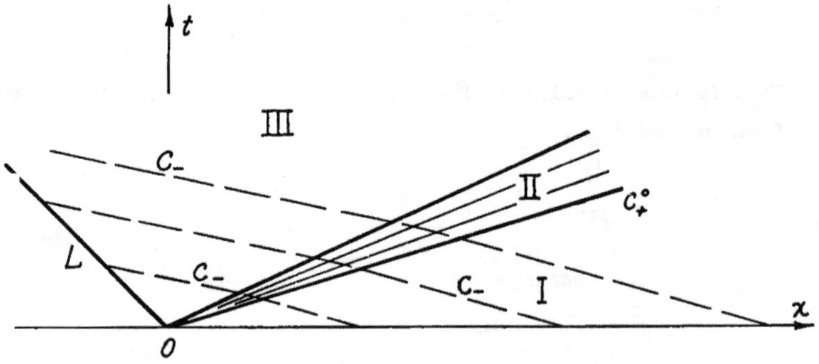

Figure 13
Non-straight characteristics
C_- in flow involving centered
rarefaction wave ($w < q_e$).

The conjugate characteristics C_- are given by the differential equation $\frac{dx}{dt} = u - c = (1 - 2\mu^2)\frac{x}{t} - 2(1 - \mu^2)c_0$ (see equations (27) and (28)) which leads, with a constant A_- of integration, to

$$(32) \qquad x = t\left\{A_- t^{-2\mu^2} - \frac{2c_0}{\gamma - 1}\right\}.$$

For $w < q_e$ the characteristics C_- emerge from the (incomplete) rarefaction zone (II) with the slope

$$(33) \qquad \frac{dx}{dt} = -w - (c_0 - \frac{\gamma - 1}{2} w) = \frac{\gamma - 3}{2} w - c_0,$$

and then continue as straight lines meeting the piston line L: $x = -wt$. For $w \geq q_e = \frac{2c_0}{\gamma - 1}$ the characteristics C_- again remain within the "complete rarefaction zone", and, since $x \sim -\frac{2c_0}{\gamma - 1}t$, they approach the particle paths asymptotically.

Obviously these considerations can be generalized to other equations of state.

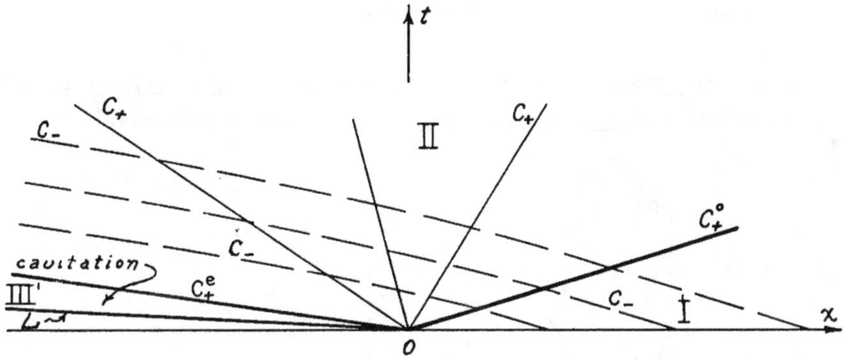

Figure 14
Characteristics C_- in flow involving centered rarefaction wave that ends in cavitation ($w > q_e$).

23. <u>Remark on centered simple waves in Lagrangean coordinates.</u> We could just as well have developed the theory of simple waves in Lagrange's coordinates, using the equations (38) or (39) of Chapter I, Art. 7. Only a brief remark is made here concerning the characteristics in the Lagrangean form. The characteristics C_+ are given by

$$\frac{dh}{dt} = \rho c = k, \tag{34}$$

and the characteristics C_- by

$$\frac{dh}{dt} = -\rho c = -k. \tag{35}$$

For centered simple waves the lines C_+ are straight lines in the h,t-plane on each of which $\frac{h}{t}$ = constant. Since $\frac{dh}{dt} = \rho c$ along a line C_+ we find $\rho c = \frac{h}{t}$. In other words, the impedance is always simply $k = \frac{h}{t}$, no matter what the equation of state is. Consequently for C_- we have $\frac{dh}{dt} = -\frac{h}{t}$, which can be immediately integrated as

$$ht = \text{constant}. \tag{36}$$

Thus, for centered rarefaction waves the non-straight characteristics are always equilateral hyperbolas in Lagrangean coordinates.

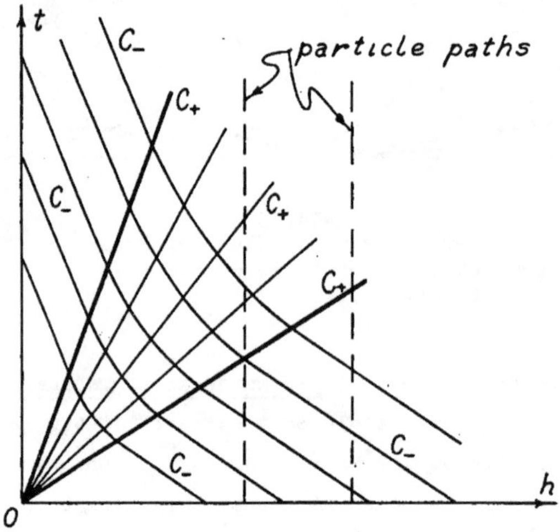

Figure 15
Characteristics in
Lagrangean coordinates
for centered simple wave.

24. Compression waves. If a piston is not withdrawn, but is moved into the gas-filled tube with a speed which never decreases, or if a receding piston is slowed down or stopped, then a contraction wave will originate at the piston. The qualitative statements and formulas pertaining to rarefaction waves also apply to contraction waves, except

that the forward characteristics C_+ no longer diverge from the piston. Density, pressure and sound speed at the piston increase, and the characteristics C_+, covering zone (II), converge and therefore have an envelope if extended sufficiently far. Certainly the solution constructed above as a simple contraction wave cannot extend beyond the envelope. For if it did, a forward characteristic C_+, with values u_A, ρ_A, corresponding to a position A of the piston, and another with values u_B, ρ_B, corresponding to a subsequent position B, would intersect beyond the envelope, and at such a point of intersection unique values of u and ρ would no longer be determined. The analytical extension of our mathematical solution beyond the envelope would therefore be multivalued and hence could no longer represent the state which occurs in reality. Physically speaking, the values of u and ρ are propagated along the characteristics C_+, and the values corresponding to a later position of the piston are propagated with a greater velocity so that they would overtake the values propagated from an earlier position of the piston. An absurdity is inescapable unless we abandon the assumption that the motion remains continuous. Consequently all compressive motion inevitably leads to discontinuities* and such discontinuities must occur before or on the envelope.

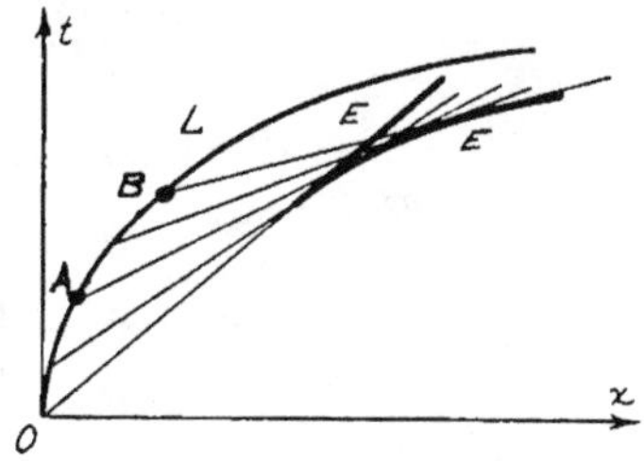

Figure 16
Contraction wave and envelope of characteristics C_+ which bounds possible location of shock line.

25. Position of envelopes for compression waves. In Section C of this chapter we shall be concerned with these discontinuities. Here a few remarks are added regarding the envelopes formed by the straight characteristics C_+ of a compression wave. Let us consider

* This was apparently first noticed by Stokes (1848); see Rayleigh [16] in the Bibliography.

an ideal gas with the adiabatic exponent γ. The piston curve L corresponding to a compression can be described by

$$L: x = f(t), \qquad f'(t) \geqq 0 .$$

Then the straight characteristics C_+ are given in terms of a parameter τ by

$$(37) \qquad x - f(\tau) = (t - \tau)\left[\frac{\gamma + 1}{2} f'(\tau) + c_0\right] ,$$

where $f'(\tau) = u_A$ is the velocity of the piston at the position A corresponding to the parameter τ. The envelope is obtained by combining the last equation with

$$(38) \quad -f'(\tau) = -\frac{\gamma + 1}{2} f'(\tau) - c_0 + \frac{\gamma + 1}{2}(t - \tau) f''(\tau) ,$$

which yields

$$(39) \quad \begin{cases} x = f(\tau) + \left\{c_0 + \frac{\gamma + 1}{2} f'(\tau)\right\} \frac{2c_0 + (\gamma - 1) f'(\tau)}{(\gamma + 1) f''(\tau)} \\ t = \tau + \frac{2c_0 + (\gamma - 1) f'(\tau)}{(\gamma + 1) f''(\tau)} \end{cases}$$

where τ ranges from zero to the value of the time for which a constant terminal speed is attained or $f''(\tau) = 0$.

As long as $f''(\tau) \geqq 0$ we have compressive motion and the envelope exists.* In the special case of a piston accelerated from rest with a constant acceleration a according to

$$L: x = \frac{a}{2} t^2, \qquad a > 0,$$

the envelope is

$$(40) \quad \begin{cases} x = \frac{\gamma}{2} a\tau^2 + \frac{2\gamma}{\gamma + 1} c_0 \tau + \frac{2}{\gamma + 1} \frac{c_0^2}{a} \\ t = \frac{2\gamma}{\gamma + 1} \tau + \frac{2}{\gamma + 1} \frac{c_0}{a} , \end{cases} \qquad \tau > 0$$

---

* For a decelerated piston, $f''(\tau) < 0$, there is no point of the envelope in the domain $x > f(t)$ corresponding to the interior of the x,t-domain of the flow.

an arc of a parabola beginning at the point

$$(41) \qquad P: \quad x' = \frac{2}{\gamma + 1} \frac{c_0^2}{a}, \qquad t' = \frac{2}{\gamma + 1} \frac{c_0}{a},$$

and tangent at this point to the characteristic line $x = c_0 t$.

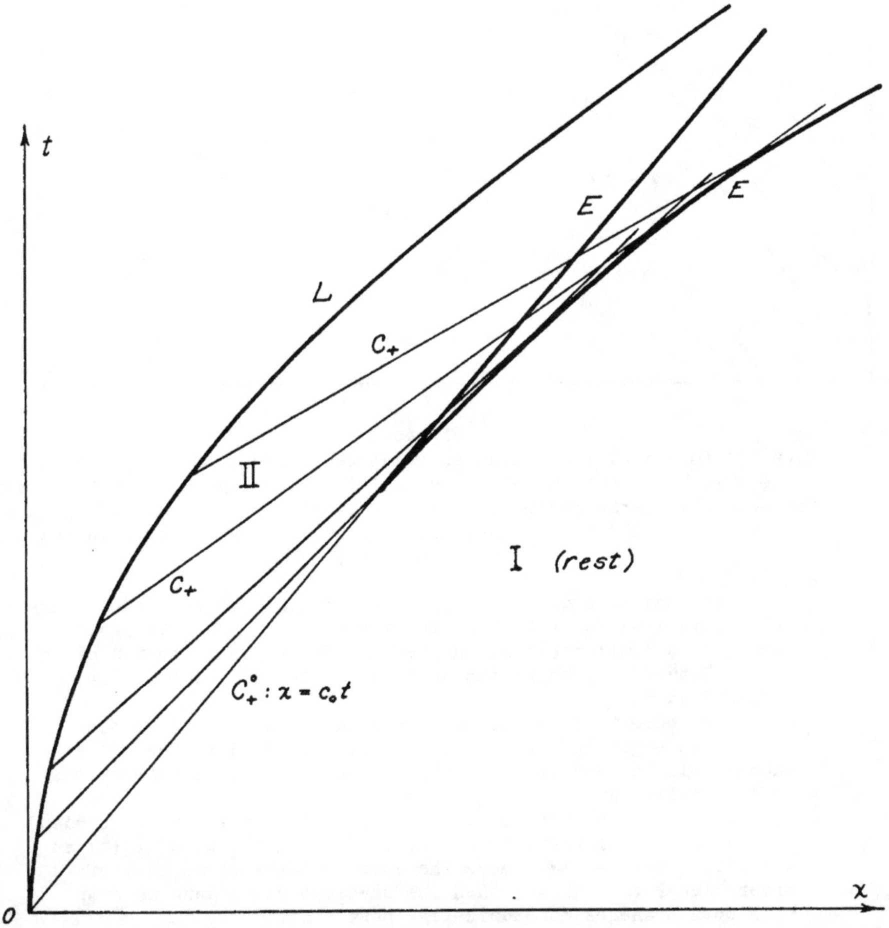

Figure 17
Envelope formed by the characteristics resulting from the piston motion $x = \frac{a}{2} t^2$, indicating the inevitable occurrence of a discontinuity.

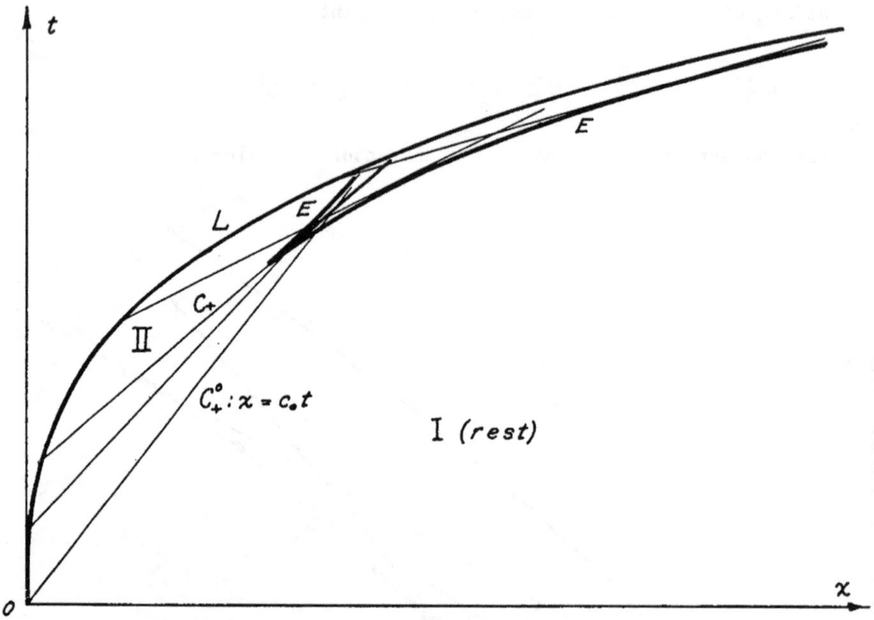

Figure 18
Envelope formed by the characteristics resulting from the piston motion $x = \frac{a}{3} t^3$. (Note that in this case the discontinuity does not begin at the boundary of the region of quiet and that the position of the envelope indicates that the position of the shock front moves closer and closer to the piston.)

We may consider the parabolic arc together with this characteristic line for $t > t'$ as forming a cusp, and it is easily seen that a similar situation prevails for a piston motion given by $x = \frac{a}{2} t^2 + \cdots$, where the dots denote terms in t of order higher than two.

We note that the coordinate x' and the time t' of the inevitable beginning of a discontinuity (later described as a shock) will be very close to the origin for large accelerations or for small sound speed c_0.

The motion just considered starts from rest with a sudden acceleration. If there is no initial discontinuity of $f''(t)$ as for $f(t) = \frac{a}{3} t^3 + \cdots$, where the dots indicate terms in t of order higher than three, then the envelope has a genuine cusp with both branches monotonically increasing.

In the typical case $f(t) = \frac{a}{3} t^3$ the envelope is given by

$$
(42) \quad \begin{cases} x = \frac{a}{3}\tau^3 + (c_0 + a\frac{\gamma + 1}{2}\tau^2)(c_0 + a\frac{\gamma - 1}{2}\tau^2)\frac{1}{a(\gamma + 1)\tau} \\ t = \tau + \frac{c_0 + a\frac{\gamma - 1}{2}\tau^2}{(\gamma + 1)a\tau} \end{cases}
$$

which shows that to $\tau = 0$ there corresponds a point at infinity on the envelope, as is always the case for $f''(0) = 0$. There will be a minimum value of t along the envelope at the point where $dt/d\tau = 0$, and since

$$
\frac{dx}{d\tau}\Big/\frac{dt}{d\tau} = \frac{dx}{dt} = c_0 + \frac{\gamma + 1}{2}f'(\tau)
$$

is finite, we see that $dx/d\tau = 0$ for the same point on the envelope. This together with the fact that $f'(\tau) > 0$ shows that the point under consideration is really a cusp as described.

Incidentally, the situation previously discussed for the piston curve $x = \frac{a}{2}t^2$ may be considered a degenerate case where the second branch of the cusp has degenerated into a straight characteristic.

The shape of the envelope may be rather complicated if the piston motion is not simple.* Since the fine features of the geometry of the envelope depend on the local behavior of the second and higher derivatives of f(t), however, we must expect that the actual behavior of the flow will not be strongly affected by geometrical complexities of the envelope. As a matter of fact, we shall see that shock discontinuities always begin infinitely weak at the cusp of the envelope. Whether they develop afterwards to strong shocks no longer depends on the local factors producing the cusp, but on the piston motion as a whole.

C. SHOCK FRONTS

26. Introduction. As we have seen, certain initial discontinuities are smoothed by centered rarefaction waves, while other motions starting as perfectly continuous contraction waves cannot be maintained

---

* One can even move the piston in such a way that the characteristics converge in a point.

without a discontinuity.* The fact is that any compression of the gas by the piston, however slow, will ultimately lead to discontinuities of velocity, pressure and density.

Hence, for a mathematical description of motions caused by impinging pistons and of many other motions as well, we must abandon, or rather supplement, the mathematical framework employed so far.

One possibility suggests itself immediately. We might try to obtain the necessary generalization from the differential equations of motion directly. In Chapter II, Art. 11, we saw that these differential equations allow discontinuities of the first and higher derivatives of u and ρ across characteristics in the u,t-plane. Such "sonic discontinuities" are associated with the differential equations; for example, they arise in initial value problems by passage to the limit from initial values with continuous derivatives to values with local discontinuities and in this limiting process the differential equations remain unchanged. In the case of linear differential equations the same type of limiting process leads to a "sonic propagation" even of discontinuities of the dependent functions themselves.** For our non-linear differential equations, however, no such sonic transmission of discontinuities of ρ and u is deducible by a passage to a limit from continuous solutions.

Hence to arrive at an adequate theory we must give up as over-simplified our original description of reality and seek a closer approximation to the actual situation by accounting for physical facts neglected in the original differential equations. Accordingly, we

---

* A particularly direct illustration of the inevitable development of discontinuities is afforded by a piston moving into a gas at rest with a speed ultimately exceeding the sound speed c_0. As shown for continuous motion, the relation $x = c_0 t$ will define a zone of rest; yet the piston itself will eventually move into the gas faster than sound, and the piston curve L will penetrate into the zone of rest. Consequently, the actual motion cannot be represented by continuous functions u and ρ of x and t.

** See Courant-Hilbert [12], Vol. II, pp. 360-361.

should introduce viscosity and heat conduction, represented by additional (linear) terms of the second order in the differential equations. To the smoothing effect of the force of viscous friction and heat conduction there corresponds the fact that, if the terms of second order are included, the differential equations have continuous solutions* no matter how small the coefficients of heat conduction and viscosity are.

Observed physical reality now points the way to a remarkable mathematical simplification. For very small values of these coefficients, the influence of heat conduction and viscosity is negligible except in the immediate vicinity of sharply defined surfaces (which may move in time) where velocity, pressure, density and temperature undergo rather sudden and large changes.** Mathematically speaking, if we let the coefficients of viscosity and heat conduction in the completed differential equations tend to zero, their continuous solution may be expected to converge to solutions of the original differential equations of first order except that certain surfaces emerge across which these solutions have discontinuities in u, ρ, p, c, T.***

The values of these quantities at both sides of such discontinuity surfaces are restricted by <u>jump conditions</u> discovered by Earnshaw, Riemann, Rankine and Hugoniot, and the decisive fact is that the effect of viscosity and heat conduction can be mathematically represented simply by these jump conditions, while otherwise the original differential equations are retained. Instead of attempting to carry out the cumbersome passage to the limit we face the simpler, though still in most cases difficult, problem of determining the surfaces of discontinuity in

---

* A mathematical proof of this statement is not attempted here.

** The fact that small viscosity and heat conductivity may safely be neglected except for discrete discontinuity layers is analogous to the boundary layer phenomenon in hydrodynamics (see Goldstein [13]).

*** A brief account of the mathematical mechanism of this limiting process will be given in Art. 34.

addition to satisfying the jump conditions and, in the regions of continuity, the original differential equations. We shall distinguish two types of such discontinuity surfaces, contact surfaces and shock fronts. The former are surfaces separating two parts of the medium without flow of substance through the surface; the shock fronts are discontinuity surfaces which are crossed by the flow of gas. If the shock front moves in time it is called a shock wave. The side of the shock front against which the flow is directed will be called the front side of the shock, the other the back side. As we shall see, the shock front, observed from the front side, always moves with supersonic speed. In this chapter we are concerned with one-dimensional motion. Hence the shock fronts and contact surfaces are assumed to be planes perpendicular to the x-axis and are represented on the x-axis by points or in the x,t-plane by lines S, henceforth called shock lines, or contact lines, respectively.

27. Shock wave in a tube. Let us first describe the simplest case of a motion involving a shock wave. The centered expansion wave caused by a piston receding at constant speed was studied as a basic type of motion. Just as basic and typical is the motion caused by a piston starting from rest and suddenly moving with constant speed w into the quiet gas. No matter how small w is, the resulting motion cannot be continuous.*

---

* For, if between the zone of quiet (I), adjacent to the x-axis, and the piston curve there were a rarefaction wave (as must happen for continuous motion), then the end of the wave nearer the piston would be the front of the wave since the flow from the piston is directed toward this end. Hence at the start of the wave the velocity of the gas would be positive and greater than the piston velocity, which is obviously absurd, since the speed of the gas there is zero. Therefore, a continuous connection of the motion of the gas at the piston with the state of rest in the zone of quiet is impossible.

What, then, will happen? Immediately there will appear a shock front moving away from the piston with a constant and, as we shall prove, supersonic speed $\dot{\xi}$, uniquely determined by the density and sound speed in the quiet gas and by the piston speed w. In front of the shock the gas is at rest, while behind the shock it moves with the constant velocity w. In the x,t-plane this very simple motion is represented in Fig. 19. The shock line S always lies in that region which would be the zone of quiet if the motion were continuous. For a sequence of decreasing values of w the shock line approaches the characteristic $x = c_0 t$ and the jump of velocity, pressure and density across the shock approaches zero. The shock becomes weak and approaches a "sonic disturbance"

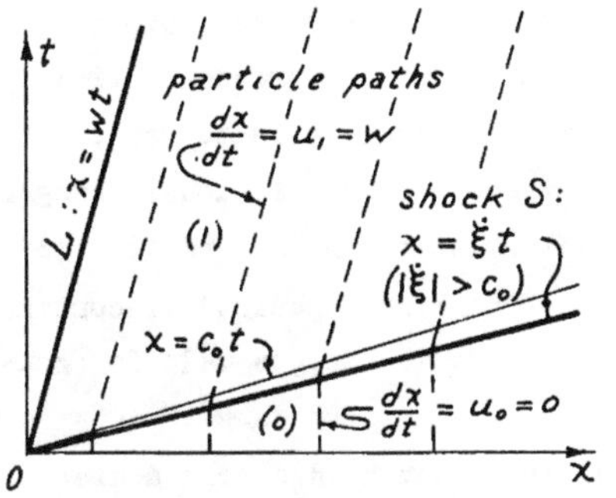

Figure 19
Shock resulting from piston moving into gas with constant velocity.

Before we can substantiate this qualitative description by proof, we must derive and discuss the jump conditions across the shock.

28. Shock conditions. We start from the following basic laws of physics:

(1°) Conservation of mass
(2°) Conservation of momentum
(3°) Conservation of energy.

Under the further assumption of continuous velocity, density and pressure, the first two laws would lead to Euler's (or Lagrange's) equations of gas dynamics. Application of these principles to discontinuous motions leads to the corresponding first two jump conditions for shocks. The energy law (3°) takes care of a more delicate point. Our original system of differential equations I(14), Art. 3, was supplemented by the equation of state in which we assumed constant entropy in keeping with the supposed adiabatic character of our processes. At first thought, it might seem plausible to suppose that even a shock discontinuity does not entail a change in entropy; in other words, that not only the continuous motion but also the shock involves merely adiabatic changes. Making this assumption, Earnshaw (1855) and Riemann (1860) considered only the conditions (1°) and (2°). However, as Rankine (1870), Rayleigh (1878) and Hugoniot (1887) observed, this procedure violates the principle of conservation of energy and thus fails to represent physical reality adequately. One must admit (discontinuous) changes of entropy across a shock and stipulate a third shock condition (3°) expressing the energy principle. This "Rankine-Hugoniot" discontinuity condition replaces the assumption of adiabatic changes made for continuous motions.

We shall now derive the conditions that hold across a discontinuity surface by applying the three general principles to a column of gas in the tube, the column covering at the time t the interval $a_0(t) < x < a_1(t)$, where $a_0(t)$ and $a_1(t)$ denote the positions of the moving particles that form the ends of the column. By e we denote the internal energy of the gas per unit mass, so that the total energy per unit mass is $e + \frac{1}{2}u^2$. Then, for the column, the three basic principles are expressed by the relations

$$(i^\circ) \qquad \frac{d}{dt}\int_{a_0(t)}^{a_1(t)} \rho\, dx = 0$$

$$(ii^\circ) \qquad \frac{d}{dt}\int_{a_0(t)}^{a_1(t)} \rho u\, dx = p_{a_0} - p_{a_1}$$

$$(iii^\circ) \qquad \frac{d}{dt}\int_{a_0(t)}^{a_1(t)} \rho\left\{\frac{1}{2}u^2 + e\right\}dx = p_{a_0}u_{a_0} - p_{a_1}u_{a_1}$$

where $p_{a_0} = p[a_0(t),t]$, etc.

Relation (i^o) needs no comment, (ii^o) expresses the fact that the rate of change of momentum of the column equals the total resulting force exerted on the column by the pressure on the two ends, (iii^o) states that the rate of increase of energy contained in the column is equal to the "power-input", i.e., the work done in unit time by the pressure against the end surfaces of the column (whose velocities are $\dot{a}_0 = u_{a_0}$ and $\dot{a}_1 = u_{a_1}$).

As long as we assume u, ρ, p continuous and differentiable in the whole column, we can easily deduce from the first two of these equations the differential equations of motion I(14), Art. 3. In the present analysis, however, we assume that in the moving column there is a point of discontinuity whose coordinate $x = \xi(t)$ moves with the velocity $\dot{\xi}(t)$.

All of our integrals have the form

$$J = \int_{a_0(t)}^{a_1(t)} \psi(x,t)dx,$$

the integrand ψ being discontinuous at $x = \xi$. Differentiation leads to

$$\frac{d}{dt}J = \frac{d}{dt}\int_{a_0(t)}^{\xi(t)} \psi(x,t)dx + \frac{d}{dt}\int_{\xi(t)}^{a_1(t)} \psi(x,t)dx$$

$$= \int_{a_0(t)}^{a_1(t)} \frac{\partial \psi(x,t)}{\partial t}dx + \left\{\psi(\xi(t)-0,t)\dot{\xi}(t) - \psi(a_0(t),t)u_0\right\}$$

$$+ \left\{\psi(a_1(t),t)u_1 - \psi(\xi(t)+0,t)\dot{\xi}(t)\right\}$$

The quantities $u_0 = \dot{a}_0(t)$ and $u_1 = \dot{a}_1(t)$ are the velocities at the ends of the column. Our formula holds no matter how short our column is, so long as it contains $x = \xi$ as an interior point. We now perform the limiting process, letting the length of the column approach zero. The first integral on the right-hand side of the last equation then tends to zero. Denoting by ψ_i, u_i the limit values of ψ, u as ξ is approached from the end a_i, we obtain immediately

$$\lim_{a_1 - a_0 \to 0} \frac{d}{dt} J = \psi_1 v_1 - \psi_0 v_0 ,$$

where

$$v_i = u_i - \dot{\xi}$$

is the flow velocity relative to the discontinuity surface. Thus we derive from our three basic equations the following conditions.

Conservation of mass

(i) $$\rho_1 v_1 - \rho_0 v_0 = 0$$

or $$\rho_0 v_0 = \rho_1 v_1 = m,$$

m being the mass flux through the surface.

Conservation of momentum

(ii') $$(\rho_1 u_1) v_1 - (\rho_0 u_0) v_0 = p_0 - p_1$$

or $$\rho_0 u_0 v_0 + p_0 = \rho_1 u_1 v_1 + p_1 .$$

By (i) this relation is equivalent to

(ii) $$\rho_0 v_0^2 + p_0 = \rho_1 v_1^2 + p_1 = P,$$

which involves only the relative velocities v.

Conservation of energy

(iii') $$\rho_1 \left\{ \frac{1}{2} u_1^2 + e_1 \right\} v_1 - \rho_0 \left\{ \frac{1}{2} u_0^2 + e_0 \right\} v_0 = p_0 u_0 - p_1 u_1$$

or $$\frac{1}{2}\rho_0 v_0 u_0^2 + \rho_0 v_0 e_0 + u_0 p_0 = \frac{1}{2}\rho_1 v_1 u_1^2 + \rho_1 v_1 e_1 + u_1 p_1 \, .$$

These relations hold across both shock fronts and contact surfaces. These two types of discontinuity surfaces were distinguished by the property that there is gas flow across a shock front, $m \neq 0$, and no gas flow across a contact surface, $m = 0$.

In this article we shall consider only shock discontinuities, postponing the discussion of contact surfaces to Art. 29.

For shocks ($m \neq 0$) relation (iii') can be simplified. Multiplying equation (ii') by $\dot{\xi}$, subtracting it from (iii'), and using (i), we obtain

$$(iii_*) \qquad \frac{1}{2} v_0^2 + e_0 + \frac{p_0}{\rho_0} = \frac{1}{2} v_1^2 + e_1 + \frac{p_1}{\rho_1} = \frac{1}{2}\hat{q}^2$$

where $\hat{q}$ is the limit speed introduced in Art. 5, Chapter I. Remembering the definition of the enthalpy $i = e + \frac{p}{\rho}$, we can write

$$(iii) \qquad \frac{1}{2} v_0^2 + i_0 = \frac{1}{2} v_1^2 + i_1 = \frac{1}{2}\hat{q}^2 \, .$$

As we see, the third shock condition has exactly the form of Bernoulli's law. It differs from it essentially, however, inasmuch as the function which represents the enthalpy i in its dependence on ρ is discontinuous across the shock, since the values i_1 and i_0 correspond to different values η_1 and η_0 of the entropy.*

The conditions (i), (ii) and (iii) represent the three shock conditions in a form in which only the relative velocities $v = u - \dot{\xi}$ are involved and not the velocities u and $\dot{\xi}$ separately. It is thus clear

---

* In other words, the change in enthalpy $i_1 - i_0$ across a shock is not equal to $\int_{(0)}^{(1)} \frac{dp}{\rho}$ but equals $\int_{(0)}^{(1)} \left(\frac{dp}{\rho} + T d\eta\right)$.

that the shock conditions are invariant under translation with constant velocity, in accordance with the Galilean principle of relativity.

Elimination of v_0 and v_1 from conditions (i), (ii), (iii_*) leads to the following important shock relation:

$$(iii_{**}) \qquad (\tau_0 - \tau_1)\frac{p_1 + p_0}{2} = e_1 - e_0 \qquad (\tau = \frac{1}{\rho}),$$

which could be interpreted to mean that the increase in internal energy across the shock front is due to the work done by the mean pressure in performing the compression. This relation is equivalent to

$$(iii_{***}) \qquad (\tau_0 + \tau_1)\frac{p_1 - p_0}{2} = i_1 - i_0 .$$

Since e, or i, is a known function of ρ and p depending on the physical properties of the gas, we have three relations between seven quantities $p_1, \rho_1, u_1, \dot{\xi}$. Hence, if three of these quantities are fixed, there is still a one-parameter family of shocks possible.

While the shock relations between the seven quantities are non-linear and thus do not necessarily define this one-parametric family uniquely, we shall see that under wide conditions, in particular, for polytropic gases*, the following theorems hold:

(A) The state (o) on the front side of the shock front and the shock velocity $\dot{\xi}$ determine the state (1) on the back side of the shock front.

(B) The state (o) and the pressure p_1 determine the shock front and the complete state (1). The same is true when, instead of p_1, the density ρ_1 or the velocity u_1 is known.

---

* See Art. 38. For liquids see Art. 33. For the general case that essentially only the condition $\frac{d^2p}{d\tau^2} > 0$ (see Art. 5, Chapter I) is imposed, see Art. 39 and the memorandum by H. Weyl [32] in which this case is treated in great detail.

Generally speaking, in view of the fact that the unknown positions of the shock lines in the x,t-plane must be determined, extremely difficult boundary value problems for our differential equations result. Many important special cases, however, are amenable to an analytical treatment as, for instance, when simple piecewise solutions of the differential equations can be fitted together across straight shock lines.

From the mathematical point of view it should be emphasized that for none of the solutions involving shock fronts has a complete uniqueness proof yet been given. Therefore, even more than is usually the case in theoretical science, the physical significance of the mathematical solutions must be verified by experiment. In gas dynamics mathematical theory is largely a means of finding qualitative and quantitative patterns which may serve to interpret experimental data.

29. Contact discontinuities. The discontinuity conditions (i), (ii'), (iii') admit of a "trivial" or degenerate solution. If the flux m through the surface of discontinuity is zero, i.e., if no substance crosses it, then we have $v_0 = v_1 = 0$, hence $u_0 = u_1 = \dot{\xi}$, and from (ii) we infer that $p_0 = p_1$, while (iii') is automatically satisfied (but (iii_*) and (iii) can no longer be deduced from (iii')). Such a discontinuity surface is called a contact surface. A contact surface moves with the gas and separates two zones of different density (and temperature); but the pressure and flow velocity are the same on both sides.* (It is obvious that in reality such a contact surface cannot be maintained for an appreciable length of time, for heat conduction between the permanently adjacent particles on either side of the discontinuity would soon make

---

* The flow velocity is continuous for one-dimensional flow. However, in flows in more than one dimension we shall consider contact surfaces across which the tangential component of flow velocity may suffer a discontinuity, while the normal component relative to the surface is zero, as in the case under discussion.

our idealized assumption unrealistic).* Henceforth, unless the contrary is stated, we shall always denote as shock only a genuine shock with the flux m through the shock front different from zero.

30. Description of shocks. We recall the following definitions given in Art. 26. The side of the shock against which the mass flux is directed was called the front side. The other side was called the back side. In other words, the particles cross the shock front from the front toward the back side. This definition is independent of the choice of coordinate system. Pressure and density, as we shall see, are always greater behind the shock than in front of it, and the degree of this increase can be used in various ways to measure the intensity of the shock (see Art. 36). Usually we shall denote the front side with the subscript ($_0$) and the back side of the shock front with ($_1$). We also say that the shock front faces the front side or is directed toward the front side.

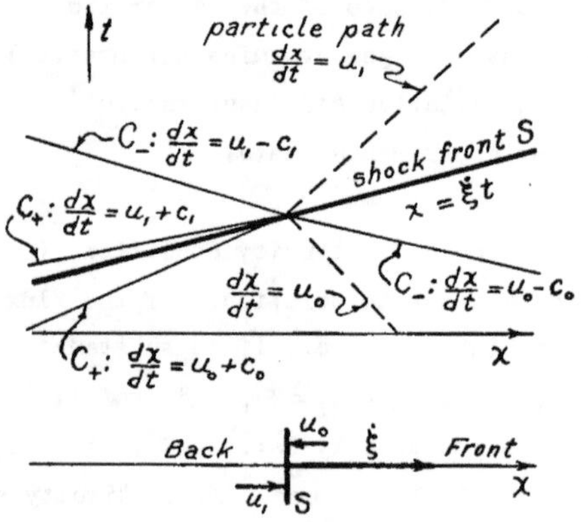

Figure 20
Shock front.

Moving shock fronts are often called shock waves and it should be clearly understood that the direction in which the shock wave moves, given by the sign of $\dot{\xi}$, has nothing to do with the direction toward which it faces, i.e., with the distinction of front and back side of the shock, the latter depending only on the relative velocity v.

We now discuss three different interpretations of a shock front,

---

* While gas particles crossing a shock front are exposed to heat conduction for only a very short time, those that remain adjacent on either side of a contact surface are exposed all the time. Hence it is clear that a contact layer will gradually spread out.

all of which are equivalent by the Galilean principle of relativity.

(a) First, suppose that the velocity u_0 on the front side is zero. Then the shock impinges on a zone (o) of rest with the velocity

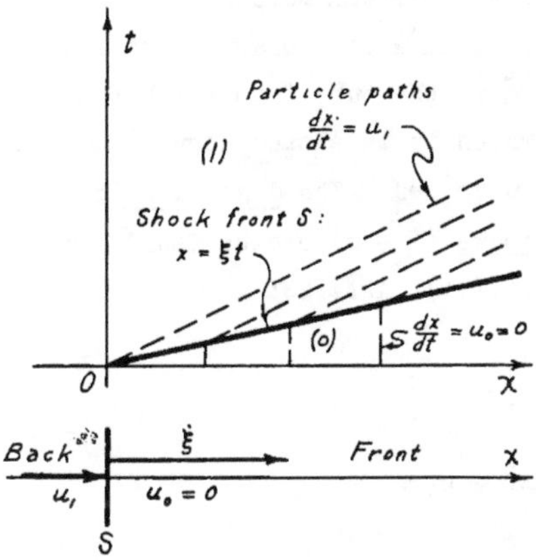

Figure 21
Impinging shock front.

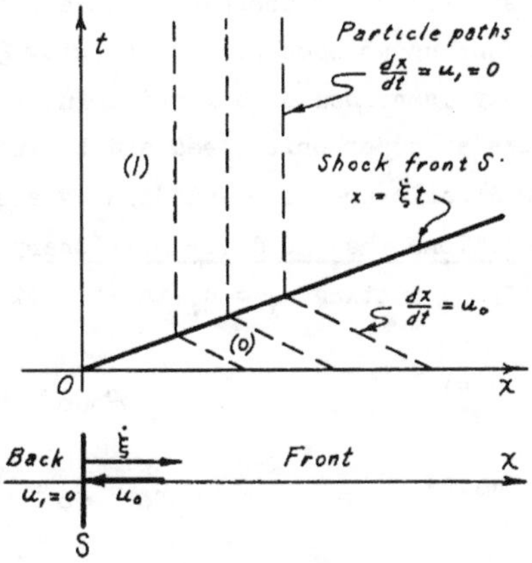

Figure 22
Receding shock front.

$\dot{\xi}$ which will be seen to be supersonic when observed from the front side, while the velocity $\dot{\xi} - u_1 = -v_1$ observed from the high-pressure zone on the back side is subsonic. The shock wave moves rapidly into the zone of quiet, enveloping more and more of the gas which, after being overtaken, follows at a speed less than that of the shock front. At the same time the density and pressure are suddenly increased.

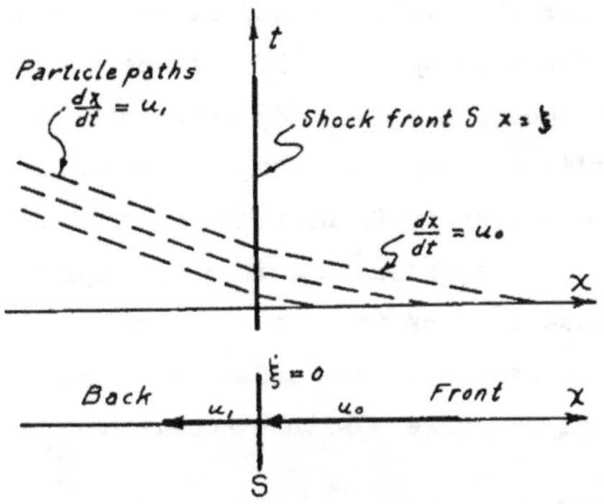

Figure 23
Stationary shock front.

(b) Secondly, suppose the velocity u_1 on the back side is zero. Then the shock front may be interpreted as receding with the velocity $\dot{\xi}$ leaving behind a high-pressure zone of quiet. Such receding shock waves will be encountered as shock waves reflected from a wall (see Art. 41).

(c) Finally, suppose that the velocity of the shock front is zero, i.e., that the shock front is stationary. (Any shock front will be stationary if observed from a coordinate system moving with the instantaneous shock front velocity $\dot{\xi}$). Such a stationary shock front is simply described by a fixed point $x = \xi$ in the tube into which the gas flows at supersonic speed and behind which it is slowed down (to subsonic speed) while pressure and density are increased. The discontinuity conditions that hold for stationary shocks ($\dot{\xi} = 0$) can be found immediately by putting $v_i = u_i$ in (i), (ii), (iii), Art. 28:

(i") $$\rho_o u_o = \rho_1 u_1 = m$$

(ii") $$\rho_o u_o^2 + p_o = \rho_1 u_1^2 + p_1 = P$$

(iii") $$\frac{1}{2} u_o^2 + i_o = \frac{1}{2} u_1^2 + i_1 = \frac{1}{2} \hat{q}^2$$

31. Models of shock motion. Shocks in their different aspects can be visualized by an analogy with a motion of particles such as a stream of fast automobiles on a highway. A stationary shock can be produced as follows. We assume a steady flow of traffic at high speed. In such a flow there will be a "sound speed", i.e., a speed at which small disturbances occurring in the traffic will spread. If the speed of the travelling cars exceeds this sound speed then a steady shock will occur at a point where the velocity is suddenly reduced, say by a change from a wide to a narrow road. There the driver in front will suddenly put on his brakes and slow down, being unable to transmit a warning signal to the driver in the rear. The increase in density is obvious and increase in pressure is also immediately represented in our model if we imagine the row of cars separated by springs or buffers with a non-linear law of repulsion.*

---

* Even an increase in temperature can be interpreted by means of such models if the energy of small vibrations of the particles is considered as representing heat.

RECEDING SHOCKWAVE

A receding shock wave can be pictured in a similar manner. Let us assume, as before, a long column of equally-spaced cars travelling at supersonic speed against an unanticipated obstacle which suddenly brings the first car to a full stop. The second will press close to the first and stop; then the third will be abruptly stopped by the second, and so on. The point separating the stopped cars from the moving cars obviously represents a receding shock front.

A shock front impinging on a zone of rest is represented by the phenomenon of a column of fast moving cars pounding against a row of widely spaced parked cars.

Models of one-dimensional wave motion by means of individual particles connected by non-linear laws of repulsion are not only suggestive, but may even be used as approximations to actual situations and thus as a basis for numerical computation.*

32. Mechanical shock conditions. We observe that only the third condition explicitly introduces the thermodynamical nature of the substance represented by the energy e or the enthalpy i as a known function of p and ρ. Hence all conclusions drawn solely from the first two shock conditions (the "mechanical conditions"),

$$\text{(i)} \qquad \rho_0 v_0 = \rho_1 v_1 = m$$

$$\text{(ii)} \qquad \rho_0 v_0^2 + p_0 = \rho_1 v_1^2 + p_1 = P,$$

are valid for any medium irrespective of its equation of state. This is true of the relations:

$$\text{(43)} \qquad m(v_1 - v_0) = p_0 - p_1$$

---

* This is proposed by von Neumann (see von Neumann [30]). It should be noted that a mechanical model as described here fails to account for a change in entropy.

(44) $$m^2 = -\frac{p_0 - p_1}{\tau_0 - \tau_1}$$

(45) $$v_1 v_0 = \frac{p_0 - p_1}{\rho_0 - \rho_1}$$

Relation (43) follows directly from (ii). Relation (44) follows from (43) by setting $v_1 = m\tau_1$ and $v_0 = m\tau_0$; relation (45) by setting $mv_1 = \rho_0 v_0 v_1$ and $mv_0 = \rho_1 v_1 v_0$.

The velocities v_0, v_1 and the mass flux m obviously have the same sign. Relation (43) then shows that the pressure p changes in the sense opposite to that of the relative speed $|v|$. Relation (44) or (45) shows that the density changes in the same sense as the pressure.

If the shock is compressive, i.e., $\rho_1 > \rho_0$, (which, as we shall see, is always the case for polytropic gases) the pressure increases and the relative speed $|v|$ decreases as the gas crosses the shock front.

Suppose that for fixed entropy η the sound speed c increases monotonically with the density ρ, and that the pressure $p(\rho,\eta)$ also increases monotonically with ρ and η; then (45) yields

$$v_0 v_1 = \frac{p(\rho_1,\eta_1) - p(\rho_0,\eta_0)}{\rho_1 - \rho_0} > \frac{p(\rho_1,\eta_0) - p(\rho_0,\eta_0)}{\rho_1 - \rho_0} = c^2(\bar{\rho},\eta_0),$$

where $\bar{\rho}$ is a properly chosen intermediate value between ρ_0 and ρ_1. Hence

$$v_0 v_1 > c^2(\rho_0,\eta_0) = c_0^2,$$

c_0 being the sound speed in the thinner medium. Since $v_0 > v_1$ (assuming $m > 0$), we have

$$v_0 > c_0,$$

i.e., the gas which has not yet reached the shock front flows with supersonic speed relative to the front. (For polytropic gases we shall presently see that after passing the shock the flow has subsonic speed relative to the front, i.e., $v_1 < c_1$).

33. Cases in which the first two shock conditions are sufficient to determine the shock. Certain further remarks should be made about the rôle of the first two shock conditions, the mechanical conditions, in contrast to that of the third, the thermodynamical condition, which is the more trenchant one. There are cases of great practical importance in which the first two conditions alone together with the pressure-density relation (equation of state) are sufficient to determine the shock phenomenon. In these cases the third shock relation remains valid, of course, but may be considered merely as a means of determining the energy balance after the problem has been solved. These remarks apply to flow in substances in which the pressure depends on the density alone and not, or not noticeably, on the entropy.

Water, for example, is approximately such a substance, inasmuch as in its equation of state, $p = A\rho^{\gamma} - B$, the coefficients A and B are approximately independent of the entropy. The internal energy for such substances splits into two parts, $e = e_1(\rho) + e_2(\eta)$, one depending only on the density, the other only on the entropy. The third shock condition can then be written

$$\left[e_2\right]_{(0)}^{(1)} = -\left[\frac{1}{2}v^2 + e_1 + \frac{p}{\rho}\right]_{(0)}^{(1)}.$$

Since the right-hand side is already determined by the first two shock conditions, $\left[e_2(\eta)\right]_{(0)}^{(1)}$ can be calculated and the rise in entropy can thus be found.

These remarks also apply to weak shocks in any substance, i.e., to shocks for which the excess pressure ratio $\frac{p_1 - p_0}{p_0}$ is small. For such weak shocks, as we shall see, the entropy rise is very small, in fact of third order in $\frac{p_1 - p_0}{p_0}$, and can therefore be safely neglected (see Art. 37).

The theorems (A) and (B) formulated earlier (Art. 28) are valid for cases where the equation of state $p = p(\rho)$ does not depend on the entropy, provided that $p'(\rho) > 0$ and $\frac{d^2p}{d\tau^2} > 0$.

To prove Theorem (A) we first show that if ρ_0 and v_0 are prescribed so that v_0 is supersonic, $|v_0| > \sqrt{p'(\rho_0)}$, then the state (1) is uniquely determined; v_1 is subsonic, $|v_1| < \sqrt{p'(\rho_1)}$. Since $\frac{d^2p}{d\tau^2} > 0$, the ratio $\frac{p - p_0}{\tau_0 - \tau}$ increases with ρ. Therefore the equation $\frac{p - p_0}{\tau_0 - \tau} = m^2$, (see (44)), has one solution ρ_1, $\tau_1 = \frac{1}{\rho_1}$, $p_1 = p(\rho_1)$. From $v_0^2 > p'(\rho_0)$ we have $m^2 > \rho_0^2 p'(\rho_0)$, which is the value that $\frac{p - p_0}{\tau_0 - \tau}$ approaches as $\rho \to \rho_0$. Hence we conclude $\rho_1 > \rho_0$, $|v_1| < |v_0|$. From $v_1^2 < v_1 v_0 = \frac{p_0 - p_1}{\rho_0 - \rho_1} < p'(\rho_1)$, (see (45)) we see that the state (1) is subsonic.

Theorem (B) is equivalent to the following statement. If ρ_0, u_0 and $\rho_1 > \rho_0$ are prescribed, states (o) and (1) are uniquely determined, provided it is in addition stipulated whether (o) should be to the left or to the right of (1). From (44) we find m^2. If (o) is to the left of (1), $m > 0$. Then $v_0 = \tau_0 m$, $\xi = u_0 - v_0$, $v_1 = \tau_1 m$ and $u_1 = v_1 + \xi$ are determined.

34. <u>Shock relations derived from the differential equations for viscous and heat-conducting fluids</u>. It seems appropriate to implement the introductory remarks in Art. 26 by a brief and somewhat more subtle analysis of how the shock conditions may be obtained by a passage to the limit of vanishing coefficients μ of viscosity and λ of heat conduction* from the differential equations involving these factors.**

(α) $\rho_t + (\rho u)_x = 0$ (Equation of continuity - same as before.)

* The notation μ, λ for these two coefficients is limited to this article only; otherwise μ^2 is used as an abbreviation of $\frac{\gamma - 1}{\gamma + 1}$.

** For derivation of these equations see Goldstein [13], Vol. II, Chapter 14.

$$(\beta) \quad \rho u_t + \rho u u_x + p_x - \frac{4}{3}\mu u_{xx} = 0$$ (Equation of motion with viscous friction.)

$$(\gamma) \quad \rho T \eta_t + \rho u T \eta_x = \frac{4}{3}\mu u_x^2 + (\lambda T_x)_x \text{ .*}$$

We consider a sudden transition in the neighborhood of a point and, with no restriction of generality, we can refer the phenomenon to a coordinate system in which this point is at rest, say at $x = 0$. For simplicity, assume that in the neighborhood of $x = 0$ the phenomenon can be considered steady, i.e., that we may set $u_t = \rho_t = \eta_t = 0$. Then we rewrite the equations by combining (α) with (β), and (α) and (β) with (γ) in the form of three conservation laws

$$(\alpha') \quad (\rho u)_x = 0$$

$$(\beta') \quad (\rho u^2 + p - \frac{4}{3}\mu u_x)_x = 0$$

$$(\gamma') \quad \left[\rho u(\frac{u^2}{2} + i) - \frac{4}{3}\mu u u_x - \lambda T_x\right]_x = 0$$

where the enthalpy i with the differential

$$di = \frac{dp}{\rho} + T d\eta$$

has been introduced. Relation (γ') represents the conservation of energy.

The possibility of writing all three of our conditions as laws

---

* The left-hand side is the heat acquired by a unit volume per unit time. The second term on the right-hand side measures the contribution due to heat conduction, while the first term measures the contribution due to viscous friction, which is essentially positive in accordance with the second law of thermodynamics.

of conservation now leads to the shock conditions in the following way. We integrate the equations (α'), (β'), (γ') between $-\varepsilon$ and ε, where ε is arbitrarily small, with the result

$$(\alpha'') \qquad \Big[\rho u\Big]_{-\varepsilon}^{\varepsilon} = 0 \; ,$$

$$(\beta'') \qquad \Big[\rho u^2 + p - \tfrac{4}{3}\mu u_x\Big]_{-\varepsilon}^{\varepsilon} = 0 \; ,$$

$$(\gamma'') \qquad \Big[\rho u(\tfrac{1}{2}u^2 + i) - \tfrac{4}{3}\mu u u_x - \lambda T_x\Big]_{-\varepsilon}^{\varepsilon} = 0 \; ,$$

in which $[f]_{-\varepsilon}^{\varepsilon}$ denotes the difference $f(\varepsilon) - f(-\varepsilon)$. For varying values of λ and μ with the limit $\lambda \to 0$, $\mu \to 0$, we consider a sequence of flows which are assumed to converge to a limit flow except possibly at the point $x = 0$.* Relations (α''), (β''), (γ''), not involving quantities at the point $x = 0$, remain valid in the limit. Thus we obtain for the limit flow

$$\Big[\rho u\Big]_{-\varepsilon}^{\varepsilon} = 0 \; ,$$

$$\Big[\rho u^2 + p\Big]_{-\varepsilon}^{\varepsilon} = 0 \; ,$$

$$\Big[\rho u(\tfrac{1}{2}u^2 + i)\Big]_{-\varepsilon}^{\varepsilon} = 0 \; .$$

When we now let ε approach zero, we obtain the same shock conditions that we found earlier.

It is clear from the procedure described that the shock conditions depend essentially on the way the differential equations were modified. If this had been done in a different manner, different shock conditions might have resulted. (If, for example, the third

---

* For a more detailed discussion of this limiting process see Becker [18] and Weyl [32].

equation were $(\eta + \lambda\phi)_x = 0$ instead of (γ'), ϕ being a function of the quantities involved, the third shock relation would have been $[\eta]_{(o)}^{(1)} = 0$, as was assumed by Riemann). This remark shows clearly that the system of shock conditions is not merely a mathematical framework essentially inherent in the unmodified differential equations but that it depends profoundly on a proper accounting for the finer features of physical reality.

It is interesting to contrast the described limit process with a different one which has always been successfully employed for linear differential equations. One considers a set of continuous solutions of the unmodified differential equations

$$(\rho u)_x = 0, \qquad (\rho u^2 + p)_x = 0, \qquad \eta_x = 0,$$

hypothetically assumed to converge to a limit solution which possibly has a discontinuity at $x = 0$. Since for the continuous solutions the relations

$$[\rho u]_{-\varepsilon}^{\varepsilon} = 0, \qquad [\rho u^2 + p]_{-\varepsilon}^{\varepsilon} = 0, \qquad [\eta]_{-\varepsilon}^{\varepsilon} = 0$$

hold, the same is true for the limit solution. Of the system of jump conditions obtained on letting $\varepsilon \to 0$, only the third differs from the shock conditions. Thus a new type of discontinuity appears to result across which the changes are adiabatic while the two mechanical conditions remain. Such a reasoning would, however, be fallacious. Since the relation $(\frac{1}{2}u^2 + i)_x = 0$ is a consequence of the three above differential equations, the relation $\left[\frac{1}{2}u^2 + i\right]_{-\varepsilon}^{\varepsilon} = 0$ also holds; and consequently the third shock condition also holds for the discontinuities of the new type. The new discontinuities would therefore be shocks in the sense discussed above but without entropy change. There are no such discontinuities, since our previous formulas imply changes* in all quantities u, p, ρ, and η across the shock front. This argument shows that continuous solutions of the unmodified differential equations never can approximate discontinuous solutions (see the remarks in Art. 1, Chapter I).

---

* See Art. 37 for change in entropy.

35. Critical speed and Prandtl's relation for polytropic gases. The third or thermodynamical shock condition becomes particularly simple in the case of polytropic gases. Then we have for the enthalpy*

$$i = \frac{\gamma}{\gamma - 1}\frac{p}{\rho} = \frac{1 - \mu^2}{2\mu^2} c^2, \quad \text{with} \quad \mu^2 = \frac{\gamma - 1}{\gamma + 1};$$

hence the condition (iii) becomes

$$(\text{iii}_\gamma) \qquad \mu^2 v_0^2 + (1 - \mu^2)c_0^2 = \mu^2 v_1^2 + (1 - \mu^2)c_1^2 = c_*^2 ,$$

where c_* is the critical speed (Art. 5, Chapter I). Due to this algebraic form of the third condition, the relations between the various quantities on both sides of the shock front and the velocity $\dot{\xi}$ of the shock front are of a purely algebraic character.

The relation between the relative velocities v_0, v_1 on both sides of the shock can be put in a very elegant and useful form, due to Prandtl, i.e.,

$$(\text{iii}_P) \qquad v_0 v_1 = c_*^2 .$$

This fundamental relation involves velocities only and does not refer explicitly to thermodynamic quantities such as pressure or density.

To prove Prandtl's relation we derive from (ii) of Art. 28, (iii_γ) and $(1 - \mu^2)\gamma = 1 + \mu^2$ the relations

$$\mu^2 P + p_1 = \mu^2 v_1^2 \rho_1 + (1 + \mu^2) p_1 = c_*^2 \rho_1 ,$$

$$\mu^2 P + p_0 = \mu^2 v_0^2 \rho_0 + (1 + \mu^2) p_0 = c_*^2 \rho_0 .$$

Subtracting, we find

$$p_1 - p_0 = c_*^2(\rho_1 - \rho_0)$$

---

* See equation I(20), Art. 4, and Art, 5, Chapter I.

or
$$\frac{p_1 - p_0}{\rho_1 - \rho_0} = c_*^2$$

and relation (iii_p) follows by (45), Art. 32.*

Prandtl's relation is evidently equivalent to the transition formula

$$(iii_p') \qquad \frac{c_*}{v_1} + \frac{v_1}{c_*} = \frac{c_*}{v_0} + \frac{v_0}{c_*}$$

($v_1 \neq v_0$ being assumed), which in turn can be proved directly as follows. From the first two shock conditions we infer that $\frac{p_1}{\rho_1 v_1} + v_1 = \frac{p_0}{\rho_0 v_0} + v_0$. This and the third condition (iii_γ) $\frac{2\gamma}{\gamma - 1}\frac{p}{\rho} = \frac{\gamma + 1}{\gamma - 1} c_*^2 - v^2$ imply that $\frac{p}{\rho v} + v = \frac{\gamma + 1}{2\gamma}\frac{c_*^2}{v} - \frac{\gamma - 1}{2\gamma} v + v = \frac{\gamma + 1}{2\gamma}\left(\frac{c_*^2}{v} + v\right)$ has the same value on both sides, whence (iii_p') results immediately.

Incidentally, Prandtl's relation exhibits the fact that if a shock is weak, i.e., if v_0 is approximately equal to v_1, then the shock is approximately a sonic disturbance, for it follows from $v_0 = v_1$ that both have the common value c_*; hence the weak discontinuity progresses approximately with sound speed relative to the gas.**

---

* Another way of obtaining Prandtl's relation is by eliminating ρ_1, p_1, i_1, ρ_0, p_0, i_0 from (i), (ii), (iii_γ) and using $c^2 = \gamma p/\rho$. The result is an equation

$$v^2 - (1 + \mu^2)\frac{P}{m} v + c_*^2 = 0,$$

which must be satisfied for $v = v_0$ as well as for $v = v_1$. As a consequence we find that $v_0 v_1 = c_*^2$.

** Of course this fact is in agreement with the principle that disturbances occurring not in the quantities u and ρ, but only in their derivatives are propagated along characteristics.

As a first, immediate consequence of Prandtl's relation we state that <u>the speed of the gas relative to the shock front is supersonic on the front side, subsonic on the back side of the shock front.</u>

For, formula (iii_P) shows that $|v_0| > |v_1|$ implies $|v_0| > c_*$ and $|v_1| < c_*$, and our assertion follows immediately from the definition of the critical speed c_*.

As a further application we derive important relations between the speed $\dot{\xi}$ of the shock front, the flow velocities u_0 and u_1 on both sides and the sound speed on one side of the shock front.

By substituting $v_i = u_i - \dot{\xi}$ in (iii_P) and using the definition of c_* we obtain

$$(46)\quad (u_0 - \dot{\xi})(u_1 - \dot{\xi}) = c_*^2 = \mu^2(u_0 - \dot{\xi})^2 + (1 - \mu^2)c_0^2$$

$$= \mu^2(u_1 - \dot{\xi})^2 + (1 - \mu^2)c_1^2 .$$

This is a quadratic equation for the shock velocity $\dot{\xi}$ and the state on one side if the velocity u_i on the other side is given. In particular, assuming that (o) is a state of rest, i.e., $u_0 = 0$, and writing $u_1 = w$, we have for $\dot{\xi}$ the equation

$$(47)\qquad (1 - \mu^2)\dot{\xi}^2 - w\dot{\xi} = (1 - \mu^2)c_0^2 ,$$

or, referring to the state (1),

$$(48)\qquad (1 - \mu^2)(\dot{\xi} - w)^2 + w(\dot{\xi} - w) = (1 - \mu^2)c_1^2 ,$$

relations that will soon prove useful.

36. <u>Relations referring to the strength of a shock for polytropic gases.</u> It is convenient for various considerations to introduce the notion of <u>shock strength</u>. Several parameters offer themselves as a measure for the strength of a shock:

the excess pressure ratio $\dfrac{p_1 - p_0}{p_0}$,

the condensation $\dfrac{\rho_1 - \rho_0}{\rho_0}$,

the parameter $M_0^2 - 1$

$M_0 = \dfrac{v_0}{c_0}$ being the Mach number of the incoming flow relative to the shock front. We write down the relations between these quantities for polytropic gases.

As stated before, (iii_{**}), Art. 28, the third shock condition can be expressed in the form

$$(\tau_0 - \tau_1)\frac{p_0 + p_1}{2} = e_1 - e_0 ,$$

which, for polytropic gases with $e = \frac{1}{\gamma - 1}\frac{p}{\rho} = \frac{1 - \mu^2}{2\mu^2}\tau p$, assumes the form $(\tau_0 - \mu^2\tau_1)p_0 = (\tau_1 - \mu^2\tau_0)p_1$. This yields the important formula

$$\frac{p_1}{p_0} = \frac{\rho_1 - \mu^2\rho_0}{\rho_0 - \mu^2\rho_1}, \tag{49}$$

which is equivalent to the relation

$$\frac{p_1 - p_0}{p_0} = \frac{1 + \mu^2}{1 - \mu^2\frac{\rho_1}{\rho_0}}\frac{\rho_1 - \rho_0}{\rho_0} = \frac{\gamma}{\left(\frac{\rho_1 - \rho_0}{\rho_0}\right)^{-1} - \frac{\gamma - 1}{2}}, \tag{50}$$

connecting the excess pressure ratio with the condensation without involving velocities.

Inversely we have

$$\frac{\rho_1}{\rho_0} = \frac{p_1 + \mu^2 p_0}{p_0 + \mu^2 p_1}, \tag{51}$$

or (52) $$\frac{\rho_1 - \rho_0}{\rho_0} = \frac{1 - \mu^2}{1 + \mu^2 \frac{p_1}{p_0}} \frac{p_1 - p_0}{p_0} = \frac{1}{\gamma\left(\frac{p_1 - p_0}{p_0}\right)^{-1} + \frac{\gamma - 1}{2}}.$$

We note that the condensation approaches the finite value $\frac{1 - \mu^2}{\mu^2} = \frac{2}{\gamma - 1}$ when the excess pressure ratio becomes infinite.

Relation (51) shows that the compression $\frac{\rho_1}{\rho_0}$ is always restricted to the range

(53) $$\mu^2 < \frac{\rho_1}{\rho_0} < \frac{1}{\mu^2};$$

so that the compression is never more than μ^{-2} - fold. For $\gamma = 1.4$ the density compression is therefore always less than 6-fold and for $\gamma = 1.2$ the limit of compression is 11-fold. For later purposes we note

(54) $$\frac{\tau_0 - \tau_1}{\tau_0} = \frac{1 - \mu^2}{\frac{p_1}{p_0} + \mu^2} \frac{p_1 - p_0}{p_0}$$

which follows from (52).

The relation between the Mach number and the excess pressure ratio is particularly simple. To derive it we substitute $\frac{v_0}{v_1}$ for $\frac{\rho_1}{\rho_0}$ in relation (49) and eliminate v_1 by Prandtl's relation $v_0 v_1 = c_*^2$. Thus we obtain the relation

$$\frac{p_1}{p_0} = \frac{v_0^2 - \mu^2 c_*^2}{c_*^2 - \mu^2 v_0^2}$$

from which, by $c_*^2 = \mu^2 v_0^2 + (1 - \mu^2) c_0^2$, we obtain

(55) $$\frac{p_1}{p_0} = (1 + \mu^2) M_0^2 - \mu^2$$

or (56) $$\frac{p_1 - p_0}{p_0} = (1 + \mu^2)(M_0^2 - 1).$$

37. Change in entropy. For polytropic gases the change in entropy $\Delta\eta$ across a shock is obtained from the expression for η given in I(5), Art. 2. We find, with c_τ denoting the specific heat at constant volume, that

$$\Delta\eta = \eta_1 - \eta_0 = \frac{1}{c_\tau}\log\left(\frac{p_1}{\rho_1^\gamma}\bigg/\frac{p_0}{\rho_0^\gamma}\right), \tag{57}$$

so that, by (49),

$$\Delta\eta = \eta_1 - \eta_0 = \frac{1}{c_\tau}\left[-\gamma\log\frac{\rho_1}{\rho_0} + \log\left(\frac{\rho_1}{\rho_0} - \mu^2\right) - \log\left(1 - \mu^2\frac{\rho_1}{\rho_0}\right)\right]. \tag{58}$$

Furthermore, for the ratio of absolute temperatures we have

$$\frac{T_1}{T_0} = \frac{\dfrac{p_1}{p_0}}{\dfrac{\rho_1}{\rho_0}} = \frac{\dfrac{\rho_1}{\rho_0} - \mu^2}{\dfrac{\rho_1}{\rho_0}\left(1 - \mu^2\dfrac{\rho_1}{\rho_0}\right)}. \tag{59}$$

With the aid of these relations we can characterize the thermodynamic changes which occur across a shock front. The equivalence of $\rho_1 > \rho_0$ and $p_1 > p_0$ follows immediately from (49) and (51). Each of these inequalities is then equivalent to $T_1 > T_0$ by (59). Now

$$\frac{d(\Delta\eta)}{d\left(\dfrac{\rho_1}{\rho_0} - 1\right)} = \frac{1}{c_\tau}\left[-\frac{\gamma}{\dfrac{\rho_1}{\rho_0}} + \frac{1}{\dfrac{\rho_1}{\rho_0} - \mu^2} + \frac{\mu^2}{1 - \mu^2\dfrac{\rho_1}{\rho_0}}\right] \tag{60}$$

$$= \frac{\gamma\mu^2}{c_\tau}\,\frac{\left(\dfrac{\rho_1}{\rho_0} - 1\right)^2}{\dfrac{\rho_1}{\rho_0}\left(\dfrac{\rho_1}{\rho_0} - \mu^2\right)\left(1 - \mu^2\dfrac{\rho_1}{\rho_0}\right)}.$$

Since $\mu^2 < 1$ and since by (53) $\mu^2 \frac{\rho_1}{\rho_0} < 1$, this derivative is positive if $\frac{\rho_1}{\rho_0} \neq 1$, i.e., $\rho_1 \neq \rho_0$. Furthermore, for $\frac{\rho_1}{\rho_0} = 1$, $\Delta\eta = 0$. The change in entropy $\Delta\eta = \eta_1 - \eta_0$ is therefore positive for $\frac{\rho_1}{\rho_0} - 1 > 0$ and negative for $\frac{\rho_1}{\rho_0} - 1 < 0$. Thus we have established that for polytropic gases <u>any one of the inequalities</u> $\rho_1 > \rho_0$, $p_1 > p_0$, $T_1 > T_0$, $\eta_1 > \eta_0$ <u>implies all the others</u>. Thus for any shock in a polytropic gas all the quantities change monotonically with $\frac{\rho_1}{\rho_0}$. As stated before, <u>shocks in polytropic gases are always compressive</u>; upon crossing the shock front <u>the gas acquires higher pressure, temperature, density and entropy</u>. This follows from the second law of thermodynamics, which stipulates that the entropy increases from the front side to the back side of the shock front.

A point of great importance is the following. The <u>change in entropy</u> across a shock front is <u>only of the third order</u> in the shock strength (i.e., in any of the quantities introduced to measure the shock strength in Art. 36). Hence, for weak shocks the jump in entropy is very small and may be neglected. Accordingly, we may treat a weak shock as an adiabatic change, and need consider only the first two shock conditions, as Riemann did in his incomplete theory (see Art. 33).

That the change in entropy $\Delta\eta$ is of <u>third order</u> in small condensation $\frac{\rho_1 - \rho_0}{\rho_0}$ can be seen immediately from (60) which shows that the derivative of $\Delta\eta$ is of <u>second order</u> in the condensation. Indeed, integration of (60) with respect to $\frac{\rho_1}{\rho_0} - 1$ leads to

$$\Delta\eta \cong \frac{1}{c_\tau} \frac{\gamma(\gamma^2 - 1)}{12} \left(\frac{\rho_1 - \rho_0}{\rho_0}\right)^3 , \tag{61}$$

where terms of higher order than three in the condensation are neglected. Using (52), this can be written in terms of the excess pressure ratio as

(62) $$\Delta\eta \cong \frac{1}{c_\tau}\frac{\gamma^2 - 1}{12\gamma^2}\left(\frac{p_1}{p_0} - 1\right)^3 ,$$

again neglecting terms of order higher than three. For shocks in water the condensation is small enough to make the assumption $\Delta\eta = 0$ reasonably valid, except when the shock is excessively strong. For gases the assumption is still justified when the shock is moderately strong.

38. The state on one side of the shock front determined by the state on the other side. The variety of relations derived for shock transitions in polytropic gases leads to simple schemes for the calculation of the state (1) when the state (o) and one additional quantity are given, and thus incidentally to a proof of the theorems A and B stated in Art. 28.

(A) Given p_0, ρ_0, u_0, $\dot{\xi}$. We find first $v_0 = u_0 - \dot{\xi}$, $c_0^2 = \gamma\frac{p_0}{\rho_0}$; then $c_*^2 = \mu^2 v_0^2 + (1 - \mu^2)c_0^2$, whereupon $v_1 = \frac{c_*^2}{v_0}$, $\rho_1 = \frac{\rho_0 v_0}{v_1}$ and

$$p_1 = p_0\left\{(1 + \mu^2)\frac{v_0^2}{c_0^2} - \mu^2\right\} \quad \text{(see (55)), or } p_1 = p_0 + \rho_0(v_0^2 - c_*^2).$$

(B) Given p_0, ρ_0, u_0 and $p_1 > p_0$. We first find $\rho_1 = \rho_0\frac{\mu^2 p_0 + p_1}{\mu^2 p_1 + p_0}$ (see (51)), then $m^2 = -\frac{p_0 - p_1}{\tau_0 - \tau_1}$ (see (44)) and choose $m > 0$ if (o) is to the left of (1); whereupon $v_0 = \frac{m}{\rho_0}$ and $v_1 = \frac{m}{\rho_1}$, and furthermore, $\dot{\xi} = u_0 - v_0$, $u_1 = v_1 + \dot{\xi}$. As a check in computation one can use

$$c_*^2 = \mu^2 v_0^2 + (1 - \mu^2)\gamma\frac{p_0}{\rho_0} = \mu^2 v_1^2 + (1 - \mu^2)\gamma\frac{p_1}{\rho_1} = \frac{p_0 - p_1}{\rho_0 - \rho_1} .$$

If the state (o) and the velocity u_1 are given, a similar procedure holds. (See details contained in Art. 33).

39. Geometric representations of shock transitions for polytropic gases. Prandtl and Busemann have given a suggestive geometric interpretation of the conditions for a stationary shock, i.e., of the algebraic relations connecting the state (o) with the state (1) adjacent to the two sides of the shock discontinuity. This representation, briefly described and supplemented here, is not restricted to a polytropic gas.

It is based on diagrams showing the dependence of the pressure p on the flow velocity u, the latter being considered as the independent variable, the quantities η and $\hat{q}$ being kept fixed; this relation is given by a curve $p = p(u)$.* If we change the value of the entropy η, keeping the value $\hat{q}$ of the limit speed fixed, we obtain a one-parametric family of such p,u-curves. These curves do not intersect; in particular, p decreases when the entropy η increases while u is fixed. This is clear from the relation $\frac{\partial}{\partial\eta}p(u,\eta) = -\rho T < 0$, which follows from $\tau dp + Td\eta = di = -udu$.

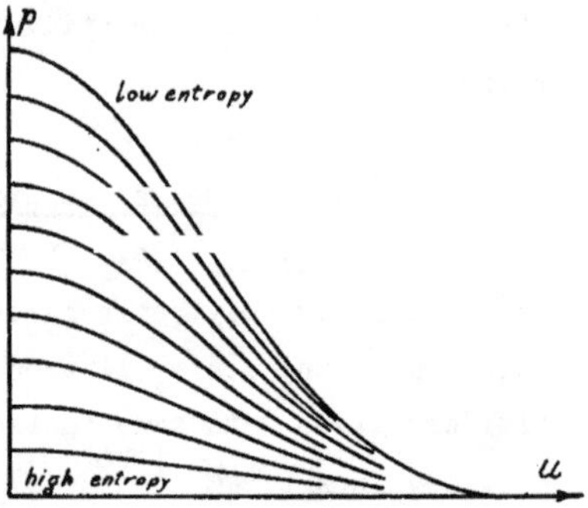

Figure 24
Pressure as a function of flow velocity u for the same limit speed and various constant values of the entropy.

Instead of u we can consider any one of the quantities i, ρ, p as independent variable in our one-parametric family of states with the same value of $\hat{q}$. Thus, from $c^2 = p_\rho$, $i = \frac{1}{2}(\hat{q}^2 - u^2)$, we find $i_u = -u$ and from $i_\rho = \frac{p_\rho}{\rho} = \frac{c^2}{\rho}$ we have $\rho_u = -\frac{\rho u}{c^2}$, hence $p_u = -\rho u = -m$

---

* Explicitly, for a polytropic gas with $p = A(\eta)\rho^\gamma$, we have

(63) $$p = A^{\frac{1}{1-\gamma}}\left[\frac{\gamma - 1}{2\gamma}(\hat{q}^2 - u^2)\right]^{\frac{\gamma}{\gamma-1}}.$$

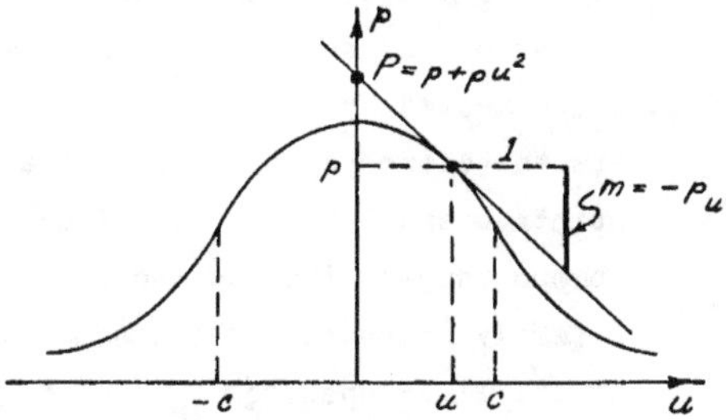

Figure 25
Graphical representation of
$m = \rho u = -p_u$ and $P = p + \rho u^2$.

and $p_{uu} = -\rho\left(1 - \frac{u^2}{c^2}\right)$. The p,u-curves therefore have negative curvature, i.e., are concave toward the u-axis, for subsonic speed, $|u| < c$, and are convex toward the u-axis for supersonic speed; hence, $u = \pm c$ are points of inflection. As was shown in Art. 5, Chapter I, under the assumptions made ($p_{\tau\tau} > 0$) there is just one value c_* (this value for polytropic gases, $c_* = \mu\hat{q}$, being independent of the entropy) such that $|u| < c$ for $|u| < c_*$ and $|u| > c$ for $|u| > c_*$. Consequently $p_{uu} < 0$ for $|u| < c_*$ and $p_{uu} > 0$ for $|u| > c_*$.

From $p_u = -\rho u$ the negative slope of our curve is seen to represent the flux crossing a unit section per unit time. The intercept of the tangent on the p-axis is therefore equal to $p + \rho u^2 = P$, a quantity appearing in the second shock condition (ii").

It is now easy to represent a shock transition graphically. Let us suppose that the shock front faces the left side (denoted by the subscript $_0$); then the flux will come from the left, i.e., we shall have $u_0 > 0$.

The third shock condition (iii") simply states that for the state (o) and the state (1) the limit speed $\hat{q}$ on both sides of the shock is the same. In our graphical representation this condition is accounted for by considering p,u-curves of the family corresponding to the same value of the limit speed $\hat{q}$ with the entropy as parameter. The two states p_0, u_0 and p_1, u_1 on the two sides of the shock front will now be represented by two points on two different curves of the family. Since $p_u\Big|_{u = u_0} = -m = p_u\Big|_{u = u_1}$, the two tangents at these points are parallel, and by virtue of the shock condition (ii") the two intercepts on the

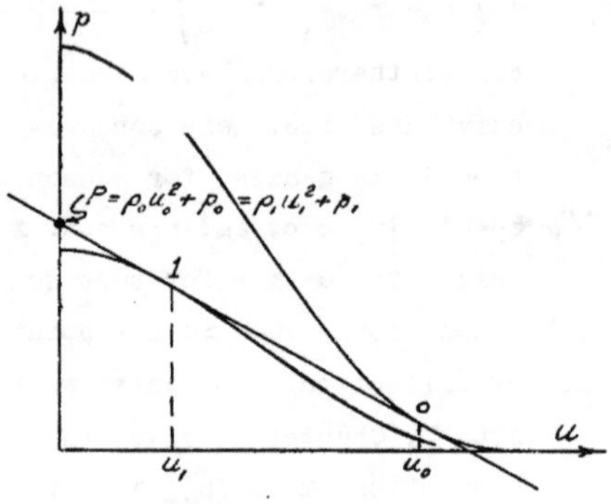

Figure 26
Representation of
shock transition
in p,u-diagram.

p-axis are equal. Hence the two tangents are identical. Any two states p_o, u_o and p_1, u_1 which can be connected by a stationary shock are therefore represented by the two points o and 1 of contact of a common tangent line to two curves of a family belonging to the same value of the limit speed $\hat{q}$.

From the figure it is clear that of the two velocities u_o, u_1 one must be supersonic, the other subsonic, since at one point $p_{uu} > 0$ and at the other $p_{uu} < 0$. Since $p_\eta < 0$ or, in other words, the p,u-curve with the greater entropy is below the other curve, it is clear that the subsonic state has greater entropy than the supersonic one.

It also follows from a continuity argument that to every supersonic state (o) there is a state (1) which can be connected with (o) by a shock.*

The preceding representation of shock transitions is not the only possible or reasonable one. Perhaps the following one, obtained by a Legendre transformation, might prove just as useful. We introduce the flux

$$m = -p_u = \rho u$$

as independent and the expression

$$P = p + mu$$

as dependent variable, and we consider the function P(m). In

---

* For a precise proof of this statement on the basis of quite different arguments see Weyl [32].

performing this transformation from p,u to P,m we must realize that by $p_{uu} \neq 0$ for $u \neq c_*$ we are assured of the feasibility of the transformation separately for $0 < |u| < c_*$ and for $c_* < |u| < \hat{q}$. Thus we obtain two branches for P(m), a "lower" branch, corresponding to $|u| > c_*$, with $P(0) = 0$, and an "upper" branch, corresponding to $|u| < c_*$, with $P(0) = P^0$. For different entropies the functions P(m)

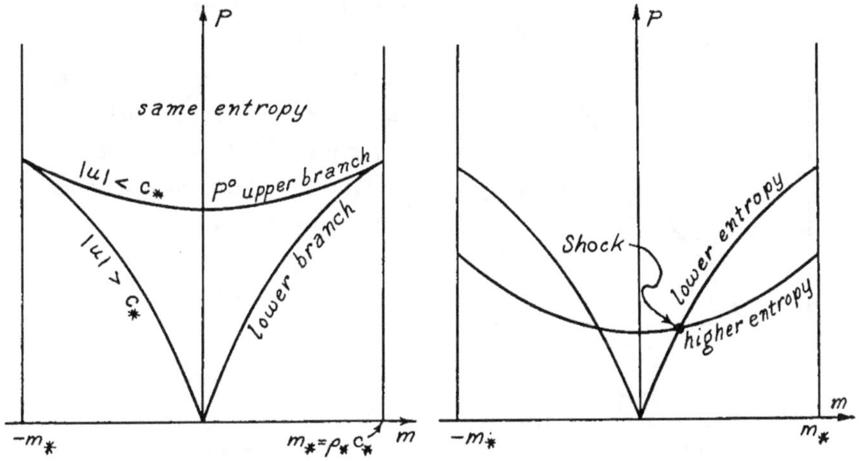

Figure 27

differ only by a constant factor. The lower branches for different entropies do not intersect each other, and the same holds for the upper branches for different entropies. However, it can be seen that through every point on a lower branch one can pass an upper branch connected with a higher entropy. Such an intersection of lower and upper branches for different entropies corresponds to a shock transition, as is obvious from our shock relations.

40. Shock conditions in Lagrangean representation. A remark concerning the Lagrangean form of the shock relations will be useful later (see Art. 7, Chapter I for definitions and notations). If x(t) is the coordinate of a moving particle, $x_0(t)$ referring to a specific "zero"-particle, then any particle is fixed (irrespective of the time) by the Lagrangean coordinate $h = \int_{x_0}^{x} \rho dx$. With h and t as independent, u and $\tau = \frac{1}{\rho}$ as dependent variables, the differential equations are (see I(39), Art. 7)

$$\tau_t = u_h, \qquad u_t = k^2\tau_h, \qquad \text{with} \qquad k = \frac{c}{\tau} = \rho c$$

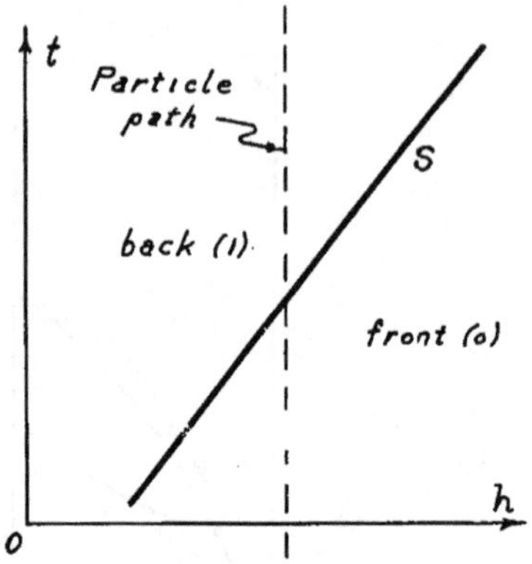

Figure 28
Motion of shock front in Lagrangean representation.

and $x_h = \tau$, $x_t = u$. Now let us consider a shock front S moving relative to the gas, enveloping at the time t a particle with the Lagrangean coordinate $h = h(t)$. Then if $x(h,t)$ is the position of the particle with the coordinates h, t, the position of the shock front is given by

$$\xi = x(h(t),t)$$

and the shock velocity is

$$\dot{\xi} = \tau\dot{h} + u.$$

With the abbreviations $\tau_1 - \tau_0 = [\tau]$, $u_1 - u_0 = [u]$ we immediately obtain the (kinematic) shock condition

$$(i_L) \qquad \dot{h}[\tau] + [u] = 0,$$

which replaces the (automatically satisfied) condition of conservation of mass. We note that $-\dot{h}$ is the mass crossing the shock front in unit time from the front side to the back side. Consequently, the conservation of momentum is expressed by

$$(ii_L) \qquad [p] - \dot{h}[u] = 0,$$

which by (i_L) in a form invariant under translatory motion is

$$[p] + \dot{h}^2[\tau] = 0,$$

while the conservation of energy, since $v = u - \dot{\xi} = -\tau\dot{h}$, is expressed by

$$\left[\frac{1}{2}(u - \dot{\xi})^2 + i\right] = 0$$

or

$$(iii_L) \qquad \frac{1}{2}\dot{h}^2[\tau^2] + [i] = 0,$$

the symbol $[f]$ always denoting $f_1 - f_0$.

41. Shock in a uniform compressive motion. The simplest basic instance of a shock transition is that between two constant states. In Art. 27 we have already given a qualitative description of the typical piston motion producing such a phenomenon. We now supply the quantitative details and proof of the mathematical consistency of our previous description. We are concerned with the problem of a piston moving with the constant velocity w into a zone (o) of rest ($u_0 = 0$) where the sound speed is c_0. Then the situation was described as follows. The piston will be preceded by a shock front S moving at supersonic speed $\dot{\xi} > c_0$ into the zone (o) of quiet. Between the impinging piston and the shock we have an ever-widening zone of those particles of the gas which the shock has suddenly accelerated from rest to the piston velocity w and which then continue to move with that velocity.

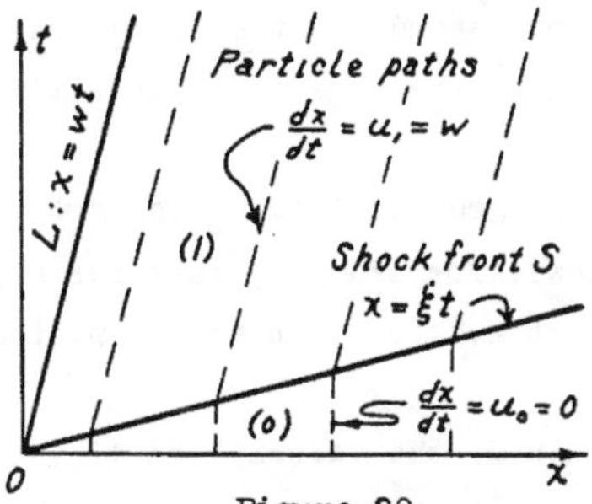

Figure 29
Shock resulting from compressive action of piston moving at a constant velocity.

To substantiate this description, we shall fit it into the framework of the preceding theory by assuming the situation described and then determining the state (1) and the shock velocity $\dot{\xi}$. Since $u_1 = w$, $u_0 = 0$, we find $\dot{\xi}$ from equation (47), obtained in Art. 35,

$$\dot{\xi}^2 - \frac{w}{1-\mu^2}\dot{\xi} - c_0^2 = 0.$$

The roots of this quadratic equation are

$$(64) \qquad \begin{cases} \dot{\xi}_+ = \dfrac{1}{2}\dfrac{w}{1-\mu^2} + \sqrt{c_0^2 + \dfrac{1}{4}\left(\dfrac{w}{1-\mu^2}\right)^2}\,, \\[2ex] \dot{\xi}_- = \dfrac{1}{2}\dfrac{w}{1-\mu^2} - \sqrt{c_0^2 + \dfrac{1}{4}\left(\dfrac{w}{1-\mu^2}\right)^2} \end{cases}$$

of which only the first is positive and corresponds to the situation we are considering. (The physical meaning of the negative root will appear in the next article).

Clearly, the shock velocity $\dot{\xi} = \dot{\xi}_+$ is greater than c_0 and greater than $\frac{w}{1-\mu^2}$. The latter observation shows, for example, that for air with $\gamma = 1.4$, $\mu^2 = \frac{1}{6}$, the shock is at least 20% faster than the piston or the oncoming column of gas.

With the shock velocity thus determined, the description of the basic compressive piston motion is shown to be consistent. Although we have given no proof of uniqueness, i.e., we have not mathematically excluded the possibility of other flow patterns, we accept the preceding reasoning as a satisfactory theory for the interpretation of actual phenomena observed under circumstances resembling our idealized model. Having obtained $\dot{\xi}$, we find by the procedure A of Art. 38 the pressure p_1 and the density ρ_1 in the zone adjacent to the piston. For high speed w of the oncoming piston (or gas), i.e., for $\frac{w}{c_0} >> 1$, we have, by (64) and (55), (51) of Art. 36,

$$(65) \qquad \dot{\xi} \sim \frac{w}{1-\mu^2} ,$$

$$(66) \qquad \frac{p_1}{p_0} \sim \frac{1+\mu^2}{(1-\mu^2)^2} \frac{w^2}{c_0^2} ,$$

$$(67) \qquad \frac{\rho_1}{\rho_0} \sim \frac{1}{\mu^2} .$$

42. Reflection of a shock on a rigid wall. We shall now discuss a point of great importance, the reflection of a shock. Suppose an oncoming column of gas of constant velocity w behind a shock front impinges on a zone of quiet bounded by a rigid wall. Then the ensuing physical phenomenon can be described as a reflection of the shock wave on the wall, and can be represented mathematically by piecewise constant solutions of the differential equations, satisfying the shock conditions across the incident shock wave and the reflected shock wave. Under the impact of the incident shock wave the zone (o) of quiet next to the wall will shrink to zero, say at t = 0; then a reflected shock will start in the opposite direction and in turn will leave a growing zone of quiet between itself and the wall. The situation can best be grasped from a diagram in the x,t-plane. State (o) is a zone of quiet characterized by the quantities $u_o = 0$, ρ_o, p_0, c_0. In the state (1) following the incident shock we have $u_1 = w$; in the state (2) adjacent to the wall we again have rest, $u_2 = 0$, but new values ρ_2, p_2, c_2. Our aim is to find the state (2) from the data ρ_0, p_0, w.

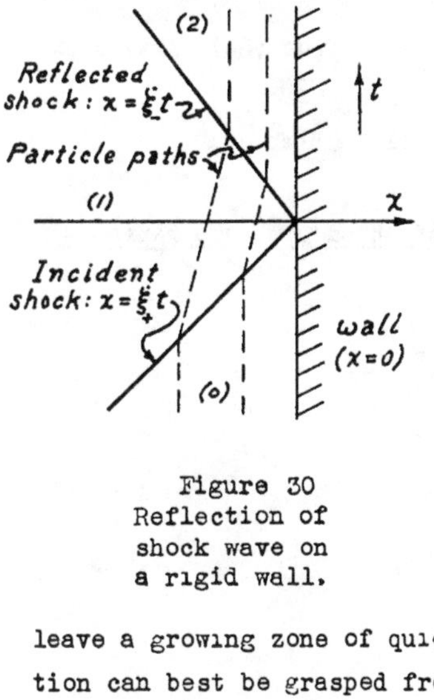

Figure 30
Reflection of shock wave on a rigid wall.

To this end we note that the pattern tentatively assumed in Fig. 30 shows a state (1) with flow velocity w and sound speed c_1 connected through a shock with a zone of rest (o) and through another shock with a zone of rest (2). $\dot{\xi}_+$ is the velocity of the incident, $\dot{\xi}_-$ the velocity of the reflected shock; then according to equation (48) of Art. 35, both these velocities must satisfy the same quadratic equation

$$(\dot{\xi} - w)^2 + \frac{(\dot{\xi} - w)w}{1 - \mu^2} - c_1^2 = 0,$$

or the two numbers $M_+ = \frac{w - \dot{\xi}_+}{c_1}$ and $M_- = \frac{w - \dot{\xi}_-}{c_1}$ satisfy the quadratic equation $M^2 - \frac{w}{(1 - \mu^2)c_1} M - 1 = 0$ so that $M_+ \cdot M_- = -1$.

Moreover, the pressure relations following from (55), Art. 36, are

$$\frac{p_o}{p_1} = (1 + \mu^2)M_+^2 - \mu^2, \qquad \frac{p_2}{p_1} = (1 + \mu^2)M_-^2 - \mu^2,$$

and, using $M_+ \cdot M_- = -1$, we obtain for the reflected pressure ratio

$$\text{(68)} \qquad \frac{p_2}{p_1} = \frac{(2\mu^2 + 1)\frac{p_1}{p_o} - \mu^2}{\mu^2 \frac{p_1}{p_o} + 1}$$

and for the excess reflected pressure ratio

$$\text{(69)} \qquad \frac{p_2}{p_1} - 1 = \frac{1 + \mu^2}{1 + \mu^2 \frac{p_1}{p_o}} \left(\frac{p_1}{p_o} - 1\right) .$$

This is the basic relation for the important phenomenon of reflection. While in linear wave motion the excess pressure after reflection is simply doubled, we find here a totally different situation. In particular, suppose we have a strong incident shock, i.e., one for which the ratio $\frac{p_1}{p_0}$ is large. We then find

$$\frac{p_2}{p_1} \sim 2 + \frac{1}{\mu^2} = \begin{cases} 8 \text{ for } \gamma = 1.4 \\ 13 \text{ for } \gamma = 1.2 \\ 23 \text{ for } \gamma = 1.1 \end{cases} \tag{70}$$

Thus, reflection of strong shocks results in considerable increase of pressure at the wall, a fact obviously of major importance.

For a weak incident shock $\frac{p_1}{p_0} - 1$ is small and we find from (69)

$$\frac{p_2}{p_1} \sim \frac{p_1}{p_0} ,$$

in agreement with the well-known facts of sonic reflection.

43. Non-uniform shocks. In the motion just discussed the situation was greatly simplified by the assumption that the shock establishes the transition from one constant state to another, implying a constant speed and strength of the shock wave. In the x,t-plane such a shock wave is represented by a straight "shock line" whose slope with respect to the t-axis is the shock velocity $\dot{\xi}$. Frequently, however, the states on the two sides of the shock front cannot both be considered constant, but are described by more complicated solutions of the differential equations. Moreover, the shock wave will not have a constant velocity, that is, the shock line in the x,t-plane will be curved. For such shocks the entropy change will in general also vary. Hence, even if the state in front of the shock is of uniform entropy, the gas, after passing the shock front, will no longer have the same entropy throughout. Then we are forced to use the more general differential equations I(14),(15),(16), Art. 3, and this is a mathematical complication which has so far precluded any complete theory, though calculations in specific cases are feasible. Fortunately, in many cases of practical importance,

the changes in entropy may be neglected with good justification (see Art. 33), and a numerical treatment of the problem becomes more feasible. In such cases we can use the simpler differential equations assuming adiabatic processes, and operate solely with the first two shock conditions disregarding the third, using the latter only for determining thermodynamical quantities after completing the solution.*

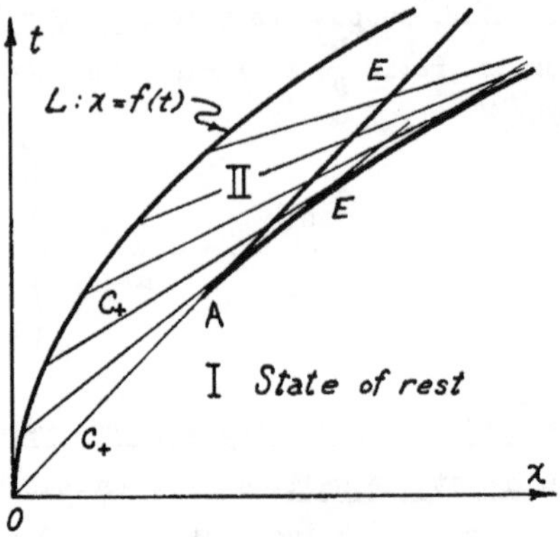

Figure 31
Simple (compression) wave and envelope of characteristics C_+ which sweep it.

The motion under the influence of a piston moving at accelerated velocity into quiet gas in a semi-infinite tube exhibits typical features of phenomena involving non-constant shocks. In an earlier discussion (Art. 24) we saw that a simple wave, represented in the x,t-plane as in Fig. 31 by a few characteristics, starts from the piston curve L: $x = f(t)$ and moves into the gas. We noticed that if the piston is accelerated, that is, if $\frac{d^2f}{dt^2} > 0$, or if $u = \dot{x}$ increases monotonically along L, then the forward characteristics C_+ starting at the piston and sweeping the domain (II) of the simple wave, have monotonically decreasing slopes $\frac{dx}{dt}$ and in general have an envelope (see the examples in Art. 25). Certainly the simple wave (II) cannot extend beyond such an envelope. We can expect the following situation to develop. From the piston curve L a simple wave (II) moves forward into the zone of quiet (I) of the gas. The envelope E of the characteristics may start after a while at a point A (in Fig. 32 this point is shown on the characteristic C_+, $x = c_0t$, through the origin, though

* This is always the case when the relation between pressure and density does not noticeably depend on the entropy (see Art. 33).

it is not necessarily situated there). Then a shock line S will begin at A. Through A we draw a characteristic C_- backwards until it meets the piston curve L at B. The curved triangle OAB represents the simple wave (II), which is completely determined by our previous theory. The shock curve S is unknown except that it lies in the region bounded by the lower and upper branches of the envelope formed by the characteristics C_+ issuing from L. This can be inferred from the facts that the converging characteristics C_+ must be cut off before they form an envelope, behind which the state would be ambiguous. Note that the motion of S is subsonic as seen from (III) and supersonic as seen from (I). Below S and OA we have a zone of quiet (I). It is the domain formed by L, AB and S which causes the deeper difficulties in the theory. Since the shock line S is not straight, the shock impresses a different entropy on the different elements of the gas crossing S from the state of rest (I) into the zone (III). In this zone, therefore, the more general differential equations I(14),(15),(16), Art. 3, are to be used. It should again be emphasized that the problem is simplified considerably whenever changes of entropy across S are negligible.

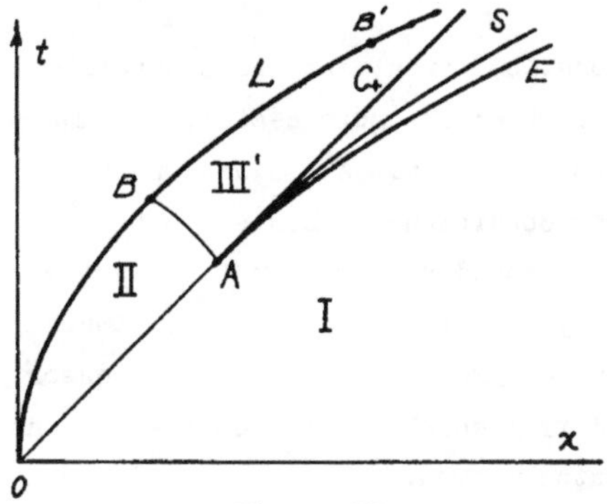

Figure 32
Compressive action of piston and resulting shock.

The shock curve S can be determined according to the following consideration. After passing the shock S from (I), we obtain definite initial values of u, ρ and the entropy η by means of the shock relations. With these initial values we solve the differential equations I(14),(15),(16), Art. 3, thereby determining u, ρ, η in a zone (III). Now the shock line S is chosen in such a way that on the arc BB' of the piston curve L this solution has the values u prescribed by the velocity of the piston.

This is a very complicated initial-boundary value problem with an unknown boundary, and no general theoretical treatment seems possible.* The reverse process, however, can be carried out more easily. Assume that we have a shock line S and determine the initial values on the other side of S according to the shock conditions. Then solve the initial value problem and find the corresponding piston motion as the motion of the flow through B. By carrying out such relatively simple computations for a suitable variety of assumed shock lines S, an assortment of flow patterns can be obtained from which one can choose the one most closely representing a given piston motion.

While determining the shock line S from a given piston motion is a difficult task, it is at least possible to analyse mathematically the very beginning of the shock, i.e., the line S in the immediate vicinity of A. This problem has been attacked by Hadamard** and more recently in an improved way by Calkin.***

It should be kept in mind that the shock is weak (sonic) at the beginning, i.e., starts with a pressure ratio $\frac{p_1}{p_0} = 1$ at the tip A. Only after the shock line S has bent away from the characteristic direction (which represents sonic disturbances) will the shock become stronger, i.e., $\frac{p_1}{p_0}$ will increase.

---

* As already mentioned (Art. 31) J. von Neumann in [30] has proposed to determine unknown shock lines by using a system of moderately few particles connected by appropriate spring forces instead of a continuous medium, and solving problems for the corresponding systems of ordinary differential equations. In preliminary tests the proposed method proved to be successful in cases where entropy changes are negligible and only the first two (mechanical) shock conditions need be considered. Whether such a procedure is practicable in case of entropy changes cannot be judged at present.

** See Hadamard, Leçons sur la Propagation des Ondes.

*** See references to a forthcoming report in von Neumann [19], p. 14.

D. Interactions

44. Introduction: General types of problems. So far we have considered primarily simple expansion, compression and shock waves originating from a state of rest, and have followed only the immediate development of the motion. What happens in reality, however, is that such waves are reflected (as already considered in Art. 42), or meet or overtake each other, so that a more general state of motion results as a consequence of various interactions in which no principle of superposition is valid. Thus, phenomena altogether different from those of linear wave motion occur, the generation of excessively high pressures by reflection of shocks being but one example.

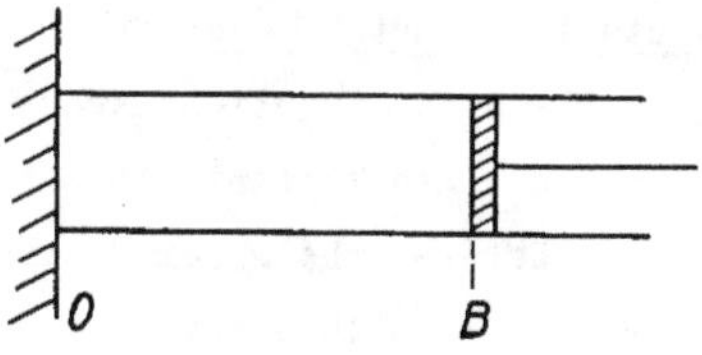

Figure 33
Initial situation in problem of Lagrange.

As a classical instance we mention the problem of Lagrange of interior ballistics. A tube is closed at a fixed point O by a rigid wall and at the other end by a piston of given mass with a variable position B. Up to time t = 0 there is atmospheric pressure in the tube. Then at t = 0 an explosion in the tube produces a gas still at rest with constant entropy η_0, density ρ_0, and a very high pressure p_0. The piston is accelerated by the pressure difference at the two sides; consequently the gas in the tube behind the piston is also accelerated and thinned out. This expansion spreads into the interior of the tube as a rarefaction wave. It travels from the piston to the wall at the point O, is reflected on the wall, meets and intersects the wave continuing from the piston into the tube, is reflected on the piston, etc. The problem is to describe the motion of the gas as well as that of the piston.*

---

* See the treatment of the problem by Love and Pidduck [17].

In this manual we shall confine ourselves to the study of what may be called elementary interactions, i.e., interactions of waves meeting contact discontinuities, or of waves facing each other and colliding, or of waves overtaking each other.

We shall see that as the ultimate outcome of interactions of these various types of wave we can in general expect two waves moving in opposite directions away from the spot of interaction.

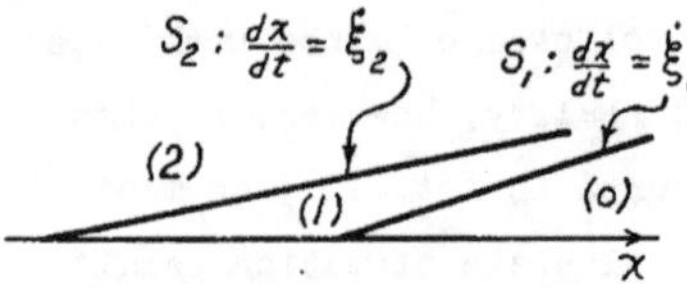

Figure 34
Overtaking of one shock by another.
$\xi_1 - u_1 < c_1$, $\xi_2 - u_1 > c_1$, i.e., $\xi_2 > \xi_1$.

A general observation concerning the important problem of overtaking of waves should be made here. Any two waves facing in the same direction, with the exception of two rarefaction waves, will eventually overtake each other.

If two shock fronts S_1 and S_2 facing in the same direction, travel one after the other, then, according to our previous result (Art. 35), the first shock S_1, observed from the region (1) between the shock fronts, travels with subsonic speed while the shock S_2 travels with supersonic speed relative to the gas in (1); hence S_2 will catch up with S_1.

Likewise, if a rarefaction wave R is followed by a shock wave S it will be overtaken (see Fig. 35), for the shock front travels faster than the sound speed in the intermediate region (1) relative to the gas in (1), while the tail of R travels with sound speed c_1 relative to (1).

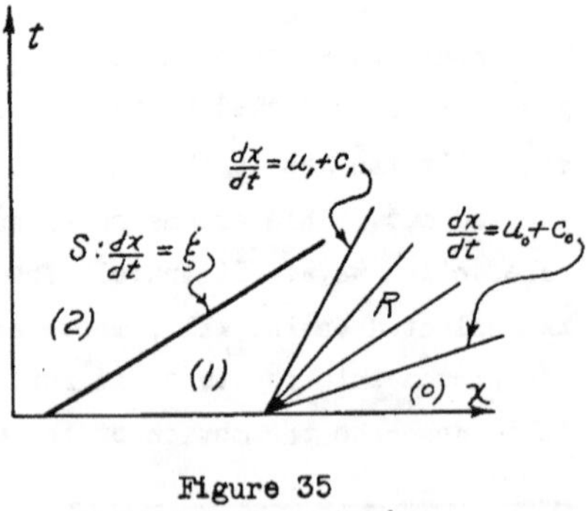

Figure 35
Shock wave overtaking rarefaction wave.
$\xi - u_1 > c_1$, i.e., $\xi > u_1 + c_1$.

In the case of a shock wave followed by a rarefaction wave (Fig. 36), the head of the rarefaction wave will travel with sound speed c_1 relative to the gas in the intermediate region (1) which is greater than the shock speed relative to the intermediate gas, since a shock front travels with subsonic speed relative to the medium behind it.

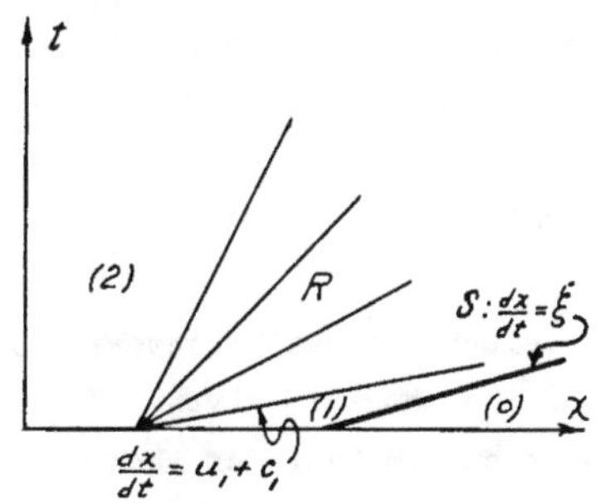

Figure 36
Rarefaction wave
overtaking shock wave.
$\dot{\xi} - u_1 < c_1$, i.e., $\dot{\xi} < u_1 + c_1$.

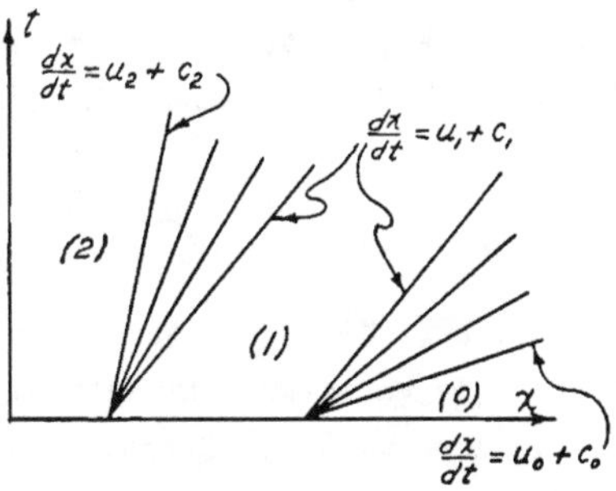

Figure 37
Two rarefaction
waves facing in
same direction.

Two rarefaction waves facing in the same direction, however, will never meet (Fig. 37). This is evident from the fact that relative to the gas in the intermediate region (1) the tail of one will travel with the same (sound) speed as the head of the other.

45. <u>Decaying shock wave</u>. The complicated motion that results when a shock wave is overtaken by a rarefaction wave and gradually diminished in strength is sometimes called a "decaying" shock wave. When the shock is weak, an approximate description of a decaying shock wave is possible (see Taylor [25], Chandrasekhar [26]). When the excess pressure ratio is less than 0.5 the change in entropy across the shock front is negligible. Therefore, the flow past the shock front may be considered as approximately isentropic. Further, as Chandrasekhar points out, the quantity $(1 - \mu^2)c - \mu^2 u$, which is constant across a forward facing simple wave (Art. 18), suffers little change across a weak forward facing shock front. Therefore, the flow at the back side of the shock front may be described as approximately a simple wave for

which

$$(71) \qquad (1-\mu^2)c - \mu^2 u = (1-\mu^2)c_0, \qquad p\rho^{-\gamma} = p_0\rho_0^{-\gamma}$$

where the subscript o refers to the state ahead of the shock front. Only one shock condition may now be used in addition to the relation (71). As such Chandrasekhar chooses the relation

$$(72) \qquad (1-\mu^2)(\dot{\xi}^2 - c_0^2) - u\dot{\xi} = 0$$

(see (47), Art. 35), from which the shock velocity $\dot{\xi}$ can be determined when the flow velocity u immediately behind the shock is given.

To construct the decaying shock wave on the basis of these simplifying assumptions, one starts with the description of the simple wave behind the shock front: u, c, ρ, p are functions of a parameter α, satisfying the equations (71), and the straight forward characteristics C_+ are given by

$$(73) \qquad C_+: \quad x = (u + c)t + x_0,$$

x_0 being also a given function of α.

Assuming $u = 0$ for $\alpha \leq 0$, which implies $\rho = \rho_0$, $p = p_0$ for $\alpha \leq 0$, one has a state of rest behind the characteristic C_+^0: $\alpha = 0$. Relation (72) determines a vector field

$$\frac{dx}{dt} = \dot{\xi}$$

through integration of which the shock line

$$x = \xi(t)$$

can be determined.

This description is, of course, only approximately correct.

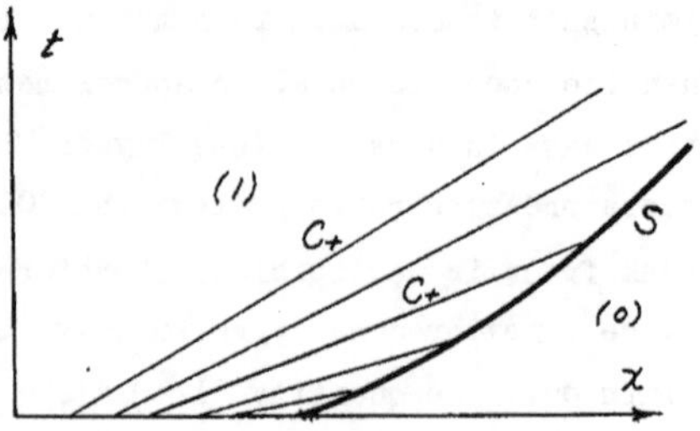

Figure 38
Decaying shock wave.

In reality, the motion behind the shock front is not a single simple wave but in addition involves a reflected wave which, however, is weak when the shock itself is weak.

It may be mentioned that Taylor's treatment is somewhat more general in that his only restriction on the change of entropy across the shock front is that it be constant. The shock transition is then determined by the two mechanical conditions. The end of the wave is placed at the characteristic on which the pressure equals the pressure ahead of the shock front. In general, the flow velocity on that characteristic and behind it turns out to be different from zero. This result seems to be in closer agreement with the discussion of later sections.

In the following, a completely different approach to the problem of interactions will be discussed.

46. Piston motion and interactions. Interactions can be studied in simple models of motion. For example, suppose that in the semi-infinite tube the piston is first moved with uniform velocity into the gas-filled tube and then is suddenly arrested. In the first phase of the piston motion a shock wave is caused; the second phase, when the piston is at rest, results in a rarefaction wave following the shock.

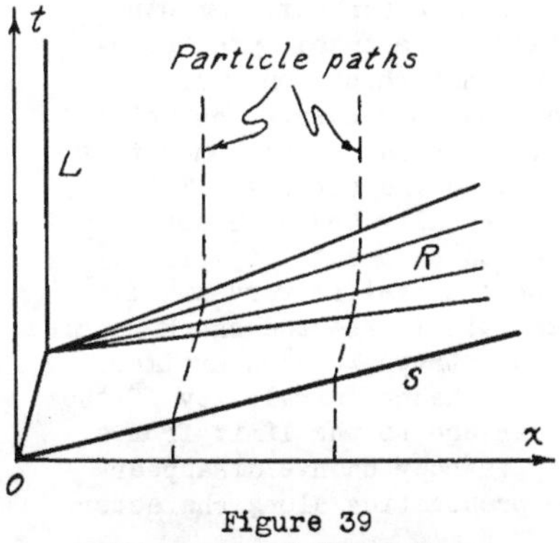

Figure 39
Motion resulting after stopping of piston moving with constant speed.

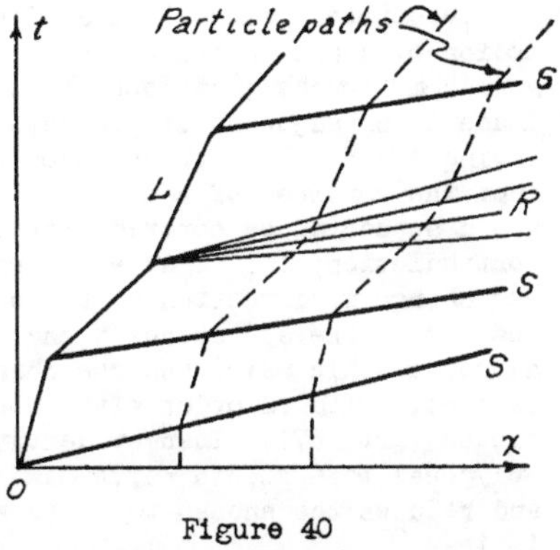

Figure 40
Motion resulting from "polygonal" piston motion.

Near the piston we obtain a motion as indicated in the diagram. However, as soon as the rarefaction wave reaches the shock front S a more complicated process starts in which the rarefaction wave is reflected as a simple wave while the shock wave becomes a decaying shock wave (see Art. 45) proceeding with diminished speed and strength.

More generally, if the piston motion consists of segments with constant speeds, changing abruptly from one constant speed to another, then the piston curve L in the x,t-plane is a more general type of polygon. From each vertex at which the piston velocity is increased a shock wave emerges, and each shock wave, as we have seen, overtakes its predecessor, i.e., the shock lines will intersect. From vertices at which the velocity decreases centered rarefaction waves start to chase the preceding shock waves. The ensuing multiple interactions lead to an exceedingly complex state in which the entropy is no longer constant throughout the gas or even over parts of it. Detailed mathematical analysis of such a flow being out of the question, we must content ourselves with a study of the elementary interactions between two elementary waves, i.e., simple waves or shock waves, or with the reflection and transmission of a wave on an interface between two media of different states.

Nevertheless, the following general consideration should not be omitted, since it clarifies a seeming paradox. Smooth motion of a piston can be approximately represented by a "polygonal" motion as just described. Now a smooth piston curve L leads to an adjacent simple wave swept by characteristics and having a constant entropy throughout, while each shock emanating from the vertices of a polygon L increases the entropy. This would appear to be contradictory, but actually there is not contradiction; for if at a vertex the sudden change in velocity (or change in direction of the polygonal curve) is very small, the shock line approaches a characteristic (since the shock becomes almost a sonic wave) and the change in entropy is even smaller, in fact, of third order with respect to change in velocity of the piston (Art. 37). Then evidently a passage to the limit from a polygonal to a smooth curve makes any entropy change disappear and reduces the shocks to sound wave propagation along characteristics.

Incidentally, the first shock condition

$$\rho_0(u_0 - \xi) = \rho_1(u_1 - \xi)$$

for a shock front emerging from a vertex of L can be written

$$\rho(u - \dot{\xi}) = (\rho + \delta\rho)(u + \delta u - \dot{\xi})$$

or

$$\delta u = \delta\rho \frac{\xi - u}{\rho + \delta\rho},$$

where $\delta u, \delta\rho$ are the increments of u snd ρ across the shock. Now if δu and $\delta\rho$ become very small and the shock approaches a sonic wave, $\dot{\xi}$ - u will approach c and in the limit we obtain

$$du = \frac{c}{\rho} d\rho ,$$

which is the basic relation for the simple wave adjacent to the smoothly moving piston.

The reflected waves which, for the nearly smooth polygonal motion L, are produced when one wave overtakes a preceding one, disappear in the limit (as can be concluded from the subsequent analysis) so that actually a simple wave results in the limit.

47. Summary of results on elementary interactions. We now give a brief account of elementary interactions, referring for details to a more complete report.* Let us first summarize the results in a qualitative way.

Interactions have very different effects according to whether the waves clash head-on or overtake each other in the same direction. (One must always bear in mind that the front of a wave is the side toward which the flow of gas is directed). The reflection of a shock wave on a rigid wall is equivalent to a special case of the head-on collision of two shock waves, namely, to the case of two symmetrical

*Courant and Friedrichs [27].

equal shock waves. In the general case of two colliding shock waves of different intensities a more complex situation results. After the two shock waves have penetrated (and thereby weakened and retarded) each other, they leave behind them an expanding zone of constant pressure and flow velocity as in the case of symmetry. In this zone, however, the density is not uniform; instead, a point moving with the flow velocity of the zone separates two regions of different (uniform) density (and temperature). In other words, a <u>contact discontinuity</u> of the type envisaged in Art. 29 appears.* This fact, which is well established experimentally, shows that we must consider contact discontinuities together with shock and rarefaction waves and study all the mutual interactions possible between any two of them.

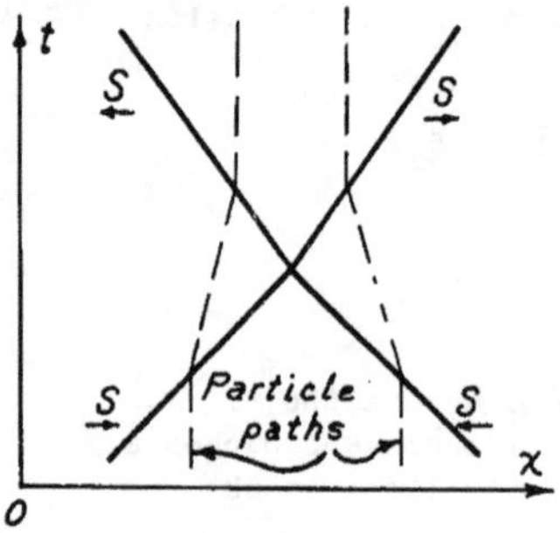

Figure 41
Head-on collision of two equal shock waves.

It is convenient to denote a shock front facing in the direction of increasing x (right shock front) by $\underset{\rightarrow}{S}$, an opposite (left) shock front by $\underset{\leftarrow}{S}$, and similarly, rarefaction waves by $\underset{\rightarrow}{R}$ or $\underset{\leftarrow}{R}$ according to whether the particles move into the wave from the right or from the left.** Contact discontinuities are denoted by the symbol T, and by the symbol TT we denote a zone in which the pressure and flow velocity are constant but in which the density, entropy and temperature vary from one particle path to another. Contact discontinuities may be distinguished as $T_>$ or $T_<$ according to whether the density is greater or smaller on the left side of T.

---

* This fact seems to have escaped writers in the field and was brought to general attention by von Neumann (see von Neumann [19]).

** Once more it should be stated that the direction in which an elementary wave <u>faces</u> has nothing to do with the direction in which the wave front <u>moves</u>.

Now the effect of <u>head-on collision of shock waves</u> can be described symbolically by the formula

$$\underset{\rightarrow}{S}\,\underset{\leftarrow}{S} \rightarrow \underset{\leftarrow}{S}\,T\,\underset{\rightarrow}{S}\,.$$

In words, head-on collisions of shock waves result in two shock fronts moving apart and separated by a contact discontinuity. Only in the case of two symmetric waves (which is equivalent to reflection of a shock on a rigid wall along the line of symmetry) will the contact discontinuity disappear.

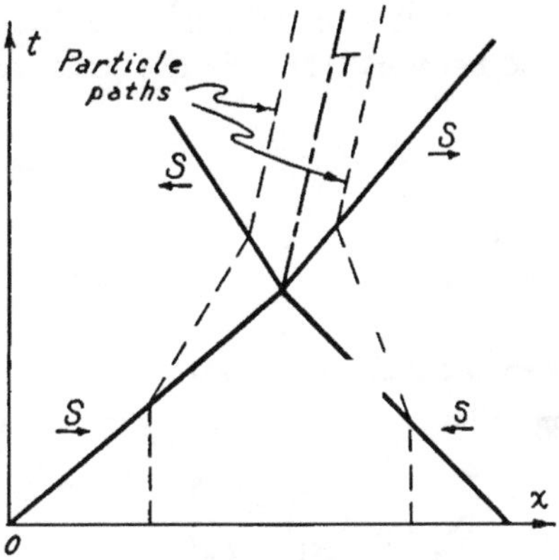

Figure 42
Head-on collision of two unequal shock waves.

<u>Overtaking of shock waves</u> in gases with an adiabatic exponent $\gamma \leq \frac{5}{3}$ results in a transmitted shock, a reflected (in general weak) rarefaction wave, and a contact discontinuity between them:*

$$\underset{\rightarrow}{S}\,\underset{\rightarrow}{S} \rightarrow \underset{\leftarrow}{R}\,T\,\underset{\rightarrow}{S} \quad \text{for} \quad \gamma \leq \frac{5}{3}\,.$$

For $\gamma > \frac{5}{3}$, which never occurs for an actual gas, we may have the same situation, but there are also cases where a shock is reflected:

$$\underset{\rightarrow}{S}\,\underset{\rightarrow}{S} \rightarrow \begin{cases} \underset{\leftarrow}{R}\,T\,\underset{\rightarrow}{S} \\ \text{or} \\ \underset{\leftarrow}{S}\,T\,\underset{\rightarrow}{S} \end{cases} \quad \text{for} \quad \gamma > \frac{5}{3}\,.$$

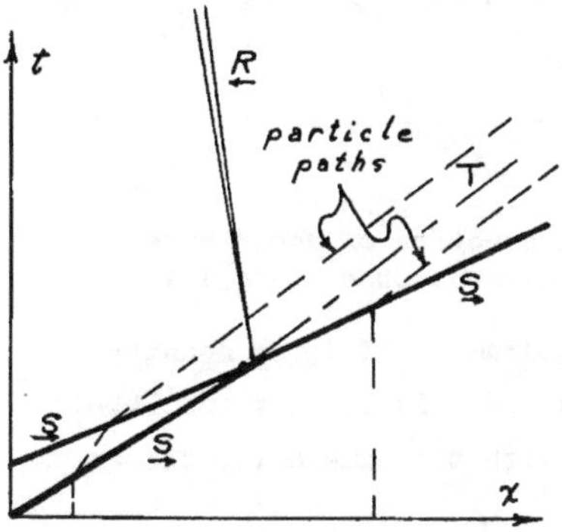

Figure 43
Overtaking of one shock wave by another.

* This result was first obtained by von Neumann [19].

Reflection and refraction of shock waves on contact surfaces (also between different media) occurs as indicated by the two formulas

$$\underset{\rightarrow}{S}\, T_< \rightarrow \underset{\leftarrow}{S}\, T_< \underset{\rightarrow}{S}\,,$$

$$\underset{\rightarrow}{S}\, T_> \rightarrow \underset{\leftarrow}{R}\, T_> \underset{\rightarrow}{S}\,,$$

whose meaning in words is that a shock wave impinging from a gas of low density on a gas of higher density results in a reflected and a

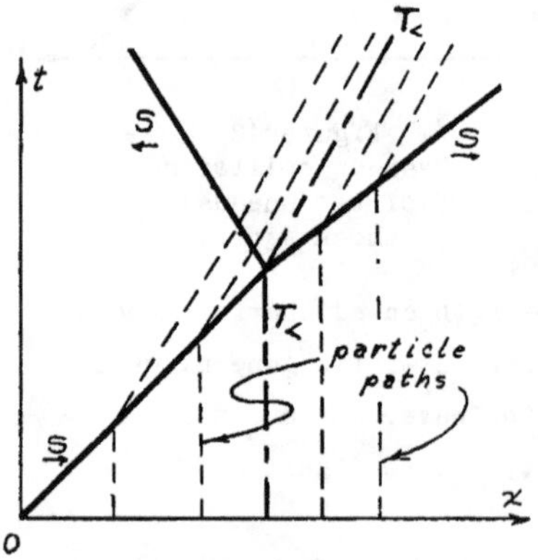

Figure 44a
Interaction of shock wave and contact surface $T_<$.

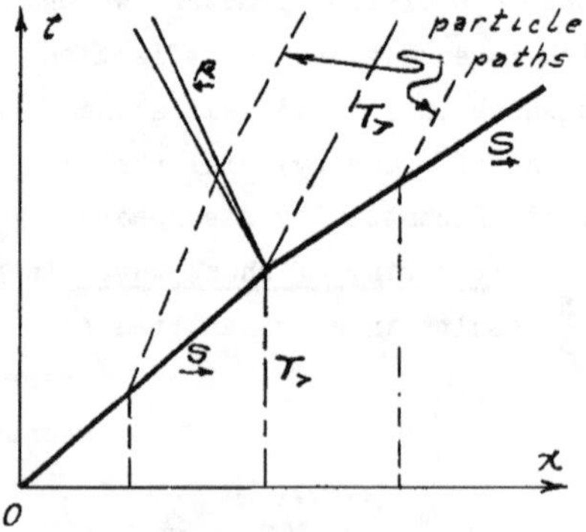

Figure 44b
Interaction of shock wave and contact surface $T_>$.

transmitted shock wave. If the second medium is of lower density a rarefaction wave is reflected but a shock wave is still transmitted. It is assumed that both media are gases with the same adiabatic exponent γ.

In interactions not involving rarefaction waves, a reflected and transmitted wave always emerge immediately after the collision. Interactions with rarefaction waves, however, at first lead to a period of penetration during which the flow cannot be described as made up of simple waves. Yet simple waves can ultimately be expected to be formed as terminal states and the following descriptions refer only to these terminal states, whose existence is assumed. For simplicity, we imagine that the process starts with two waves separating zones of constant pressure and flow velocity. (In a sufficiently small neighborhood of the collision this is no essential restriction of generality). Then, interactions with rarefaction waves lead to terminal states as follows. The head-on collision of two rarefaction waves (which in cases of symmetry is equivalent to reflection of a rarefaction wave on a rigid wall) again yields two rarefaction waves as the terminal state:

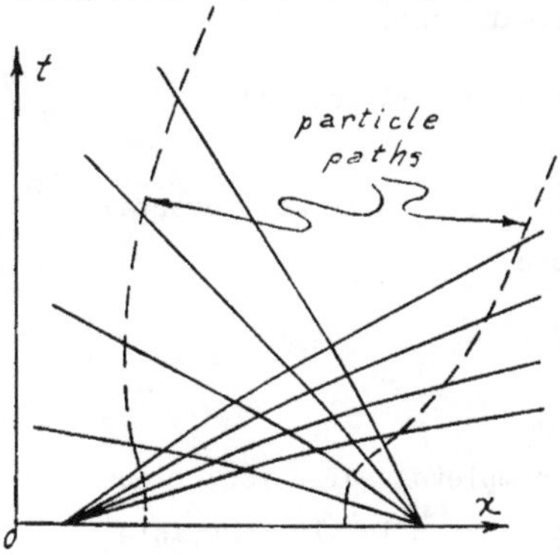

Figure 45
Head-on collision of two rarefaction waves.

$$\underset{\rightarrow}{R}\,\underset{\leftarrow}{R} \rightarrow \underset{\leftarrow}{R}\,\underset{\rightarrow}{R} \;.$$

Likewise, the outcome of a <u>collision of a rarefaction wave with a zone of higher density</u> is simply described by

$$\underset{\rightarrow}{R}\,T_< \rightarrow \underset{\leftarrow}{R}\,T_<\underset{\rightarrow}{R}$$

while the interaction

$$\underset{\rightarrow}{R}\,T_>$$

is essentially more complicated (see Courant-Friedrichs [37]).

<u>A shock wave overtaking a rarefaction wave</u> can result in three different types of terminal states, depending on whether the shock is strong enough to consume the whole rarefaction wave and to cross it:

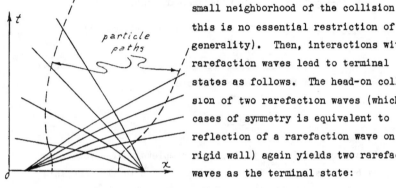

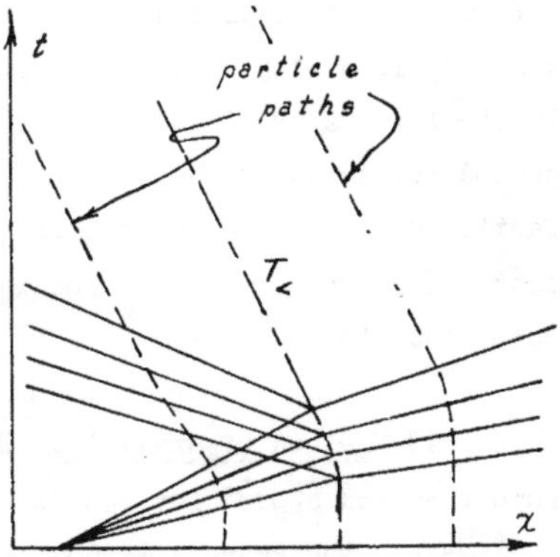

Figure 46
Interaction of rarefaction wave and contact surface $T_<$.

$$\underset{\rightarrow}{S}\,\underset{\rightarrow}{R} \rightarrow \underset{\leftarrow}{S}\;TT\;\underset{\rightarrow}{S} \qquad \text{(for strong oncoming shock)}$$

$$\underset{\rightarrow}{S}\,\underset{\rightarrow}{R} \rightarrow \underset{\leftarrow}{S}\;TT\;\underset{\rightarrow}{R} \qquad \text{(for weak oncoming shock)}$$

$$\underset{\rightarrow}{S}\,\underset{\rightarrow}{R} \rightarrow \underset{\leftarrow}{S}\;TT \qquad \text{(for an intermediate situation).}$$

The case of a decaying shock wave, i.e., a shock wave overtaken and gradually devoured by a rarefaction wave

$$\underset{\rightarrow}{R}\,\underset{\rightarrow}{S}$$

has been discussed earlier (Art. 44) by a completely different kind of analysis but this important type of interaction is also amenable to the treatment employed in the cases now under discussion (see [27],p.46).

Two further comments relating to these descriptions and figures are necessary. Zones TT result from shocks penetrating into a rarefaction wave (with the effect of mutual weakening); the shock line is bent and the particles crossing the shock undergo different changes in entropy. That ultimately all these particles emerge with the same velocity is a simplifying assumption which has been found to be true in the first approximation.* This assumption, however, cannot be expected to represent the facts accurately. In various cases a rarefaction wave or a shock wave is reflected as a compression wave and the latter in turn leads to a shock. Our schematic formulas refer only to such an ultimate shock.

48. Method of analysis. While reference must be made to the report quoted above** for the detailed justification of the preceding statements, the general method for obtaining these results can be

---

* In an unpublished computation carried out by AMP-NYU.

** Courant and Friedrichs [27].

indicated here. It amounts to an algebraic discussion of the transition relations for shock fronts and rarefaction waves, guided by a graphical representation.

By subscripts a, b, m, ℓ, r, k we denote zones of constant values of p and u, the letters ℓ and r meaning "left" and "right" or smaller and larger values of x respectively. It is convenient to represent states in a p,u-plane. In such a representation two constant states p,u between which shock and rarefaction waves take place

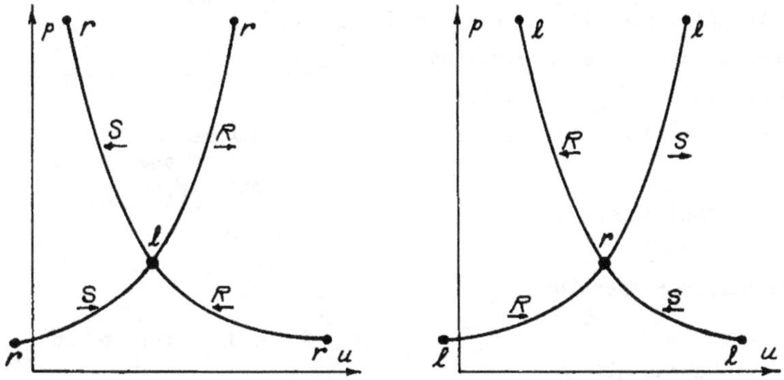

Figure 47
Diagrams representing graphically the states (u_ℓ, p_ℓ) and (u_r, p_r) on left and right sides respectively of shock and rarefactions waves.

can then be represented as in the accompanying diagrams (Fig. 47). For shocks we always have $u_r < u_\ell$; for rarefaction waves $u_r > u_\ell$ irrespective of the directions in which these waves face. The quotient $\frac{p_r - p_\ell}{u_r - u_\ell}$ is always positive for forward waves and negative for backward waves.

From our previous analysis we recall and infer the following results. If a state (a) of the gas $u = u_a$, $p = p_a$, $\rho_a = \frac{1}{\tau_a}$, and another state (b), with the quantities u_b, p_b, $\frac{1}{\tau_b}$, are connected through a shock wave, then from (43), (44), Art. 32, and (54), Art. 36, we have $\frac{p_b - p_a}{u_b - u_a} = m = \pm\sqrt{\frac{p_b - p_a}{\tau_a - \tau_b}} = \pm\sqrt{\frac{p_b + \mu^2 p_a}{(1-\mu^2)\tau_a}}$. Hence

(74) $u_b = u_a \pm \phi_a(p_b)$

where

(75) $\phi_a(p) = (p - p_a)\sqrt{\dfrac{(1-\mu^2)\tau_a}{p + \mu^2 p_a}}$,

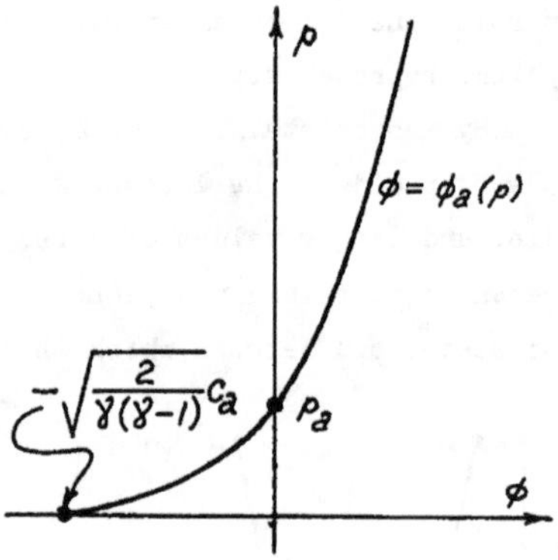

Figure 48
Graph of the function $\phi_a(p)$.

and where, as will be seen immediately, the plus sign is to be taken for shock fronts $\overrightarrow{S}$, the minus sign for shock fronts $\overleftarrow{S}$. The monotonic function ϕ, incidentally, satisfies

(76) $\phi_a(p_b) = -\phi_b(p_a)$,

has the simple properties

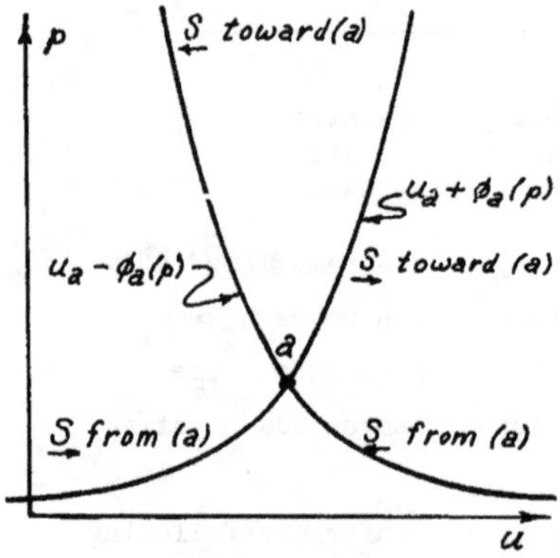

Figure 49
Locus of all states which can be connected with a given state (a) through shock waves $\overrightarrow{S}$ and $\overleftarrow{S}$ facing toward and away from (a) as indicated.

(77) $\begin{cases} \phi_a(p) \uparrow \infty & \text{for } p \uparrow \infty, \\ \phi_a'(p) \downarrow 0 & \text{for } p \uparrow \infty, \end{cases}$

and touches the ϕ_a-axis at

$$\phi_a(0) = -\sqrt{\frac{2}{\gamma(\gamma - 1)}}\, c_a$$

where c_a is the sound speed in region (a). Relations (74) are shown graphically in Fig. 49; that the various branches of the curves correspond to shock waves facing toward or away from (a) as indicated may be seen immediately from the consideration that $u_r < u_\ell$ and the pressure behind the shock wave exceeds that in front.

In Fig. 50 the loci of all states which can be connected by shock waves $\underset{\rightarrow}{S}$ and $\underset{\leftarrow}{S}$ with a given state (r) to the right or (ℓ) to the left, are indicated.

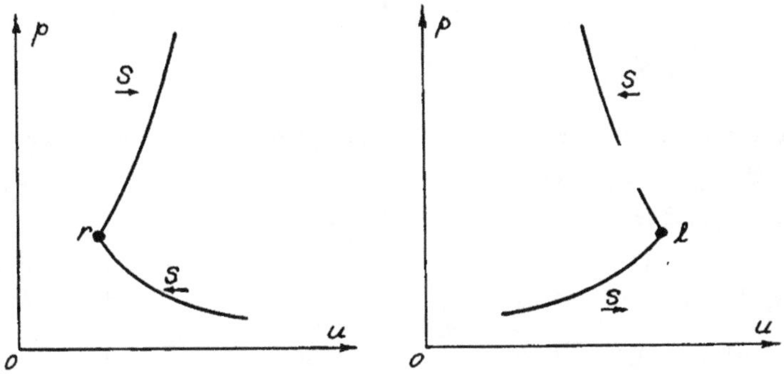

Figure 50

Locus of possible states (ℓ) on left of a shock wave if state (r) on right is prescribed.

Locus of possible states (r) on right of a shock wave if state (ℓ) on left is prescribed.

A similar representation can be obtained for the one-parametric family of constant states (b) which can be connected with a fixed state (a) by a rarefaction wave. We saw (Art. 18) that across a rarefaction wave $\underset{\rightarrow}{R}$ or $\underset{\leftarrow}{R}$, $u \mp \frac{2}{\gamma - 1} c = u \mp \frac{2}{\gamma - 1}\sqrt{\gamma \tau p} = \text{constant}$. Now, since only adiabatic changes of state occur, we have $\frac{\rho_a}{\rho_b} = \left(\frac{p_a}{p_b}\right)^{\frac{1}{\gamma}}$ or $\sqrt{\tau_b p_b}\, p_b^{-\frac{\gamma - 1}{2\gamma}} = \sqrt{\tau_a p_a}\, p_a^{-\frac{\gamma - 1}{2\gamma}}$. Thus

$$u_b - u_a = \pm \frac{2}{\gamma - 1}\left(\sqrt{\gamma \tau_a p_a} - \sqrt{\gamma \tau_b p_b}\right)$$

$$= \pm \frac{\sqrt{1 - \mu^4}}{\mu^2}\, \tau_a^{\frac{1}{2}} p_a^{\frac{1}{2\gamma}}\left(p_a^{\frac{\gamma - 1}{2\gamma}} - p_b^{\frac{\gamma - 1}{2\gamma}}\right)$$

so that, analogous to (74), we have

(79) $$u_b = u_a \pm \psi_a(p_b)$$

where

(80) $$\psi_a(p) = \frac{\sqrt{1-\mu^4}}{\mu^2}\,\tau_a^{\frac{1}{2}}\,p_a^{\frac{1}{2\gamma}}\left(p^{\frac{\gamma-1}{2\gamma}} - p_a^{\frac{\gamma-1}{2\gamma}}\right)$$

and where the plus sign prevails for waves $\overrightarrow{R}$ and the minus sign for waves $\overleftarrow{R}$. The monotonic function ψ has the properties

(81) $$\begin{cases} \psi_a(p) \uparrow \infty & \text{for } p \uparrow \infty \\ \psi_a'(p) \downarrow 0 & \text{for } p \uparrow \infty, \end{cases}$$

touches the ψ_a-axis tangentially at

$$\psi_a(0) = -\frac{\sqrt{1-\mu^4}}{\mu^2}\sqrt{\tau_a p_a}$$

$$= -\frac{1-\mu^2}{\mu^2}c_a,$$

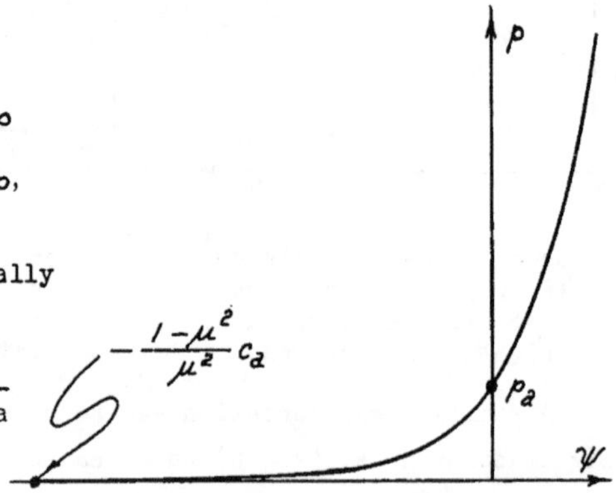

Figure 51
Graph of the
function $\psi_a(p)$.

and satisfies the relation

(82) $$\psi_a(p_b) = -\psi_b(p_a).$$

As in the case of shock waves, and for similar reasons, the possible states (b) as given by (79), which can be connected with a fixed state (a) through rarefaction waves facing toward and away from (a), are shown graphically in Fig. 52. In Fig. 53 are shown the loci of all states which can be connected with a given state (ℓ) or (r), on the left or right of the wave respectively, by rarefaction waves $\overrightarrow{R}$ or $\overleftarrow{R}$.

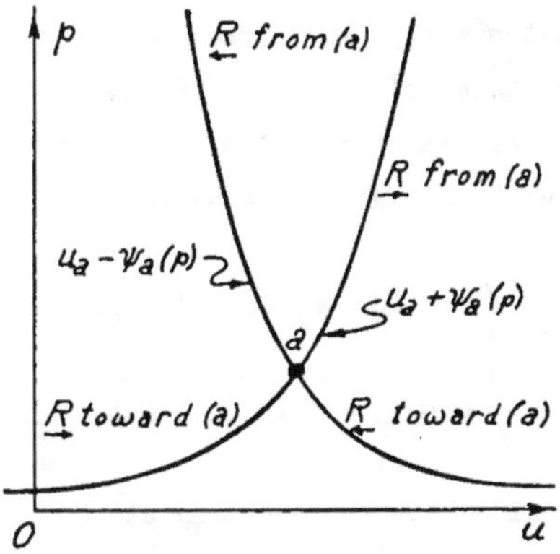

Figure 52
Locus of all states which can be connected with a state (a) through rarefaction waves $\overrightarrow{R}$ and $\overleftarrow{R}$ facing toward or away from (a) as indicated.

Now the problem of interactions can be attacked as follows.* Suppose that before the interaction we have a state (ℓ) to the left and a state (r) to the right, both joined to the middle zone (m) by (known) waves. At the instant the interaction begins, the middle zone (m) disappears and, either instantaneously or after a period of penetration, a forward wave moves into the state (r) and a backward wave moves into the zone of state (ℓ), the two latter waves now being separated by a new middle zone (m^*) of constant velocity u^* and constant pressure p^*. All that we have to find is the state (m^*) and the waves connecting (r) and (ℓ) with (m^*). See Fig. 54.

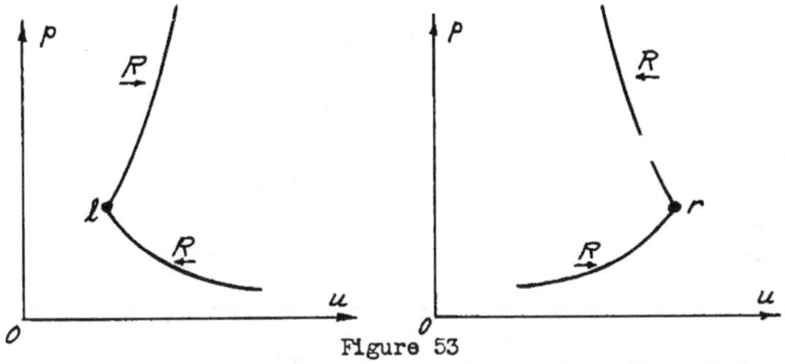

Figure 53

Locus of possible states (r) on right of a rarefaction wave if state (ℓ) on left is prescribed.

Locus of possible states (ℓ) on left of a rarefaction wave if state (r) on right is prescribed.

---

* The same method applies to Riemann's problem where at t = 0 two arbitrarily prescribed states u_1, ρ_1, p_1 and u_0, ρ_0, p_0 collide at a point, and the resulting motion is to be determined.. The slight difference with respect to the problem discussed above is that the two given

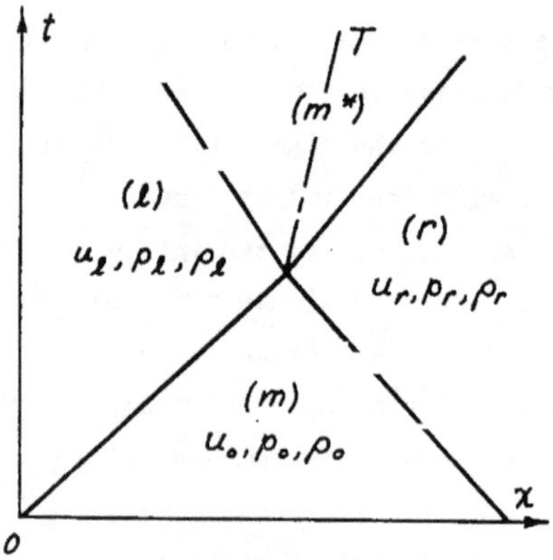

Figure 54
Showing middle zone (m) between two colliding waves (in this case shock waves) and middle zone (m^*) separating waves resulting after collision.

To this end we note that all states which can be reached from a state (r) by a forward wave must be represented in the p,u-plane by a curve Γ_r expressed analytically by

$$(83)\quad \Gamma_r \begin{cases} u = u_r + \phi_r(p), & p > p_r, \quad \underset{\rightarrow}{S}, \\ u = u_r + \psi_r(p), & p < p_r, \quad \underset{\rightarrow}{R}, \end{cases}$$

while all points representing states connected with (ℓ) by a backward wave are on a curve

$$(84)\quad \Gamma_\ell \begin{cases} u = u - \phi_\ell(p), & p > p_\ell, \quad \underset{\leftarrow}{S}, \\ u = u - \psi_\ell(p), & p < p_\ell, \quad \underset{\leftarrow}{R}. \end{cases}$$

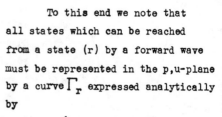

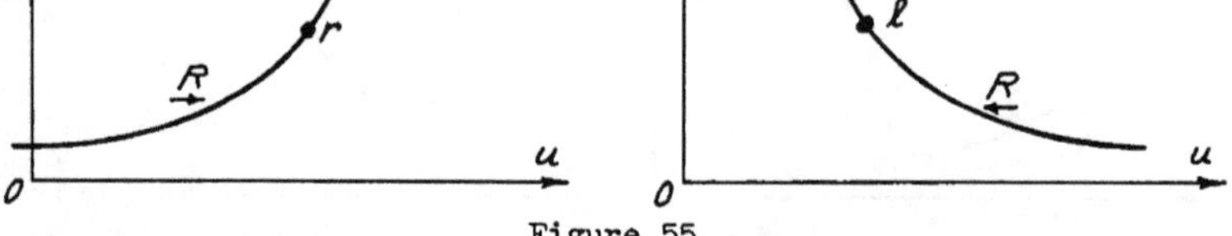

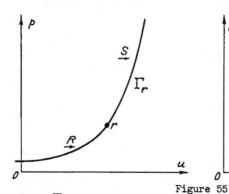

Figure 55

Curve Γ_r representing all states which can be reached from a given state (r) on the right of forward waves $\underset{\rightarrow}{S}$ or $\underset{\rightarrow}{R}$.

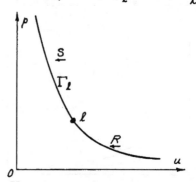

Curve Γ_ℓ representing all states which can be reached from a given state (ℓ) on the left of backward waves $\underset{\leftarrow}{S}$ or $\underset{\leftarrow}{R}$.

---

states are here considered independent, while only five of their data can be arbitrarily prescribed if they both are connected by shocks with a previous intermediate state (m) (which has shrunk to a point by the time t = 0.

If the two states (ℓ) and (r) are known we need simply draw the two Γ curves through them. The point of intersection then determines the state (m^*) and the waves from (m^*) to (ℓ) and to (r). In practice, graphical construction often will not be sufficiently precise, but it indicates the proper arrangement for numerical calculation.

To illustrate the procedure we consider a few cases in more detail.

(a) To study the clash of two shock waves $\underset{\rightarrow}{S}$ and $\underset{\leftarrow}{S}$ we observe that our states (ℓ) and (r) must be represented in the p,u-diagram as in Fig. 56 since they are obtained from (m) by a forward and backward shock respectively (and are therefore not entirely independent, a fact which is not essential to our procedure). The curves Γ_r and Γ_ℓ, according to our diagram, must intersect in a point m^*, and m^* is on the upper part of Γ_r as well as Γ_ℓ. Hence the two transitions from m^* are shocks, as stated.

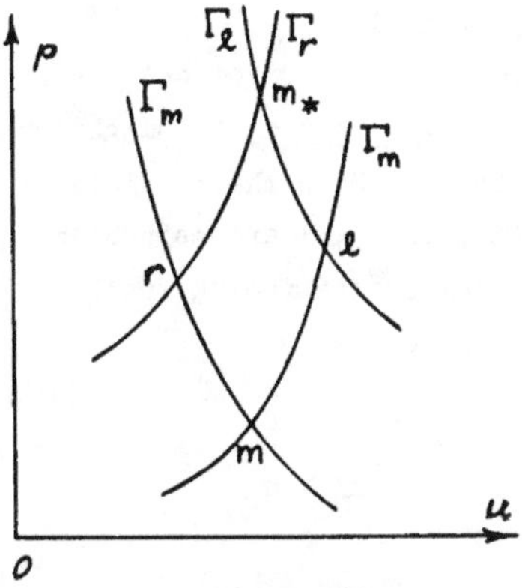

Figure 56
Graphical analysis of head-on collision of two shock waves. (See also Fig. 42).

In the state (m^*) we have constant values of p and u. The shock transition from (r) to (m^*), however, in general determines a value ρ_+^* different from the value ρ_-^* obtained by the shock transition from (ℓ). Hence, in the zone (m^*) of the x,t-plane, a line of density discontinuity coinciding with the particle path from the point of collision.

(b) The collision between two rarefaction waves leads to a terminal state which can be determined just as easily. Here the relative positions of (m), (ℓ) and (r) and consequently of (m^*) is immediately seen to be that* of Fig. 57. But this shows that (m^*) is on those parts of Γ_r and Γ_ℓ to which rarefaction waves correspond. Hence our previous statement $\underset{\rightarrow}{R}\,\underset{\leftarrow}{R} \rightarrow \underset{\leftarrow}{R}\,\underset{\rightarrow}{R}$ is justified.

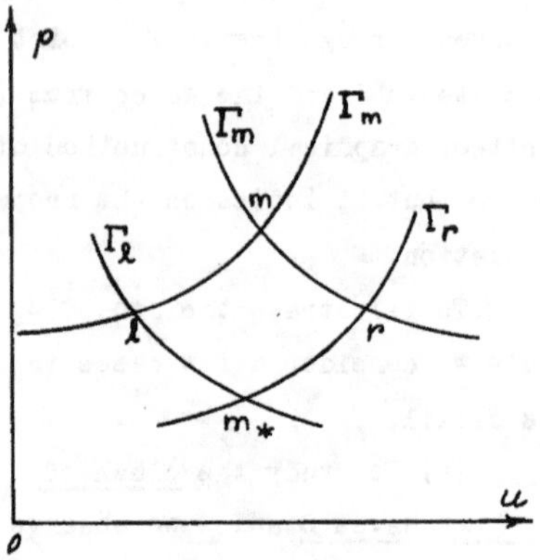

Figure 57
Collision of two rarefaction waves (see also Fig. 46).

(c) The problem of <u>one shock wave overtaking another</u> is slightly more delicate. Here the original states (ℓ), (m), (r) are separated by two shock fronts facing to the right; we have

$$p_\ell > p_m > p_r$$
$$u_\ell > u_m > u_r \; .$$

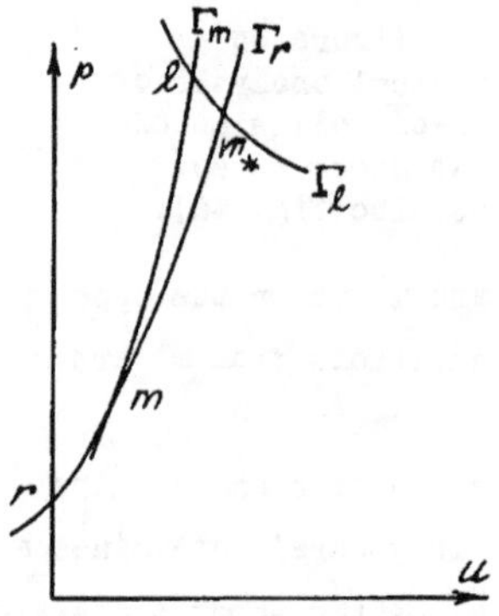

Figure 58
One shock wave overtaking another (see also Fig. 43).

In the p,u-diagram the point m is evidently on the curve Γ_r since m is connected to r by a shock and the unknown point m^* must be situated on the same Γ_r. On the other hand, m^* lies on a Γ_ℓ which connects it with ℓ and hence a situation as in Fig. 58 results if Γ_r passes on the right side of ℓ. It is possible to show that this is the case from the algebraic form of our functions ϕ and ψ if $\gamma \leq \frac{5}{3}$.** Then our diagram indicates that an intensified forward shock $\underset{\rightarrow}{S}$ (between r and m^*) and a weak backward

---

* See Courant-Friedrichs [27], p. 28, for the case where no point of intersection m^* exists.

** See Ibid., Appendix A3.

rarefaction wave result. For the same reason as in (a) we can expect a contact discontinuity in zone (m^*).

If $\gamma > \frac{5}{3}$, however, situations are possible where Γ_r passes to the left of $\mathcal{L}$, so that we obtain a (weak) reflected shock instead of a reflected rarefaction wave.

49. The process of penetration. When rarefaction waves are involved in interactions, first a more complicated flow occurs before two waves emerge and move apart. This process of penetration requires the solution of non-trivial boundary value problems for our differential equations of flow. As observed before, one even has to consider the more general system I(14),(15),(16), Art. 3, in which non-uniform shocks, and thus variable entropy and transition bands TT, are involved.

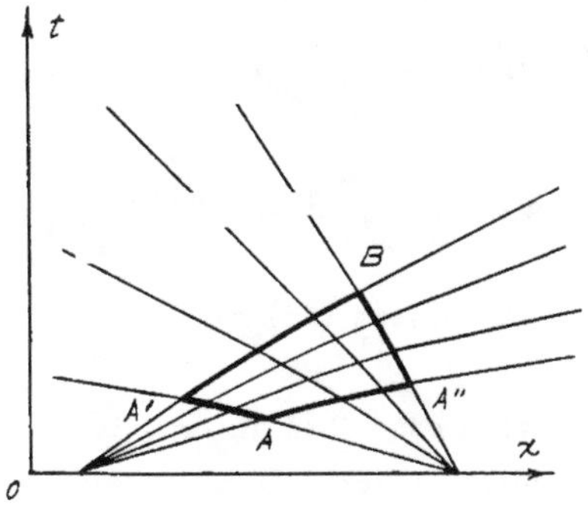

Figure 59
Interaction of two rarefaction waves showing zone AA'BA'' of penetration.

Hardly any mathematical work has been done in these cases*, except for the collision of two rarefaction waves. In this case the problem depends on the simpler equations I(14),(17), Art. 3, and can easily be formulated.

Suppose the two waves $\underset{\rightarrow}{R}$ and $\underset{\leftarrow}{R}$ are given rarefaction waves. Assume that they are centered, although it is not necessary to do so. Then we know not only the straight characteristics sweeping these waves, but also the curved conjugate characteristics. In particular, through the first point of collision we know the two characteristics AA' and AA'' (Fig. 59) closing off the oncoming simple waves. Now we also know the value of u and ρ on the characteristic angular line A'AA'', and the problem is to determine

---

* See, however, v. Neumann [30], and first footnote at end of Art. 43.

u and ρ in a quadrangle A'AA"B formed by four characteristics, of which the two sides A'B and A"B have to be determined together with the solution.

This is a so-called characteristic initial value problem. Its solution constitutes an important part of Riemann's classical paper. Riemann linearizes the problem by interchanging the rôles of dependent and independent variables; then the machanism of "Riemann's method of integration" of hyperbolic differential equations can be applied with the help of Riemann's explicit construction of "Riemann's function" in terms of hypergeometric functions. In our presentation the situation can be understood by introducing characteristic coordinates β and α (see Art. 9, Chapter II), or, more conveniently, certain functions r and s of β and α (see Art. 18, Chapter III). Then our initial value problem can be shown to refer to the equations

(85)
$$x_s - (u + c)t_s = 0 \qquad 2r = u + \int_{\rho'}^{\rho} \frac{c}{\rho}\, d\rho$$
where
$$x_r - (u - c)t_r = 0 \qquad 2s = u - \int_{\rho'}^{\rho} \frac{c}{\rho}\, d\rho$$

Since $u \pm c$ are known functions of r and s, the system (85) may be linearized (see p. 23).

However, by using the method of iteration described in Art. 10, Chapter II, we can deal directly with this characteristic initial value problem for the characteristic system corresponding to the equations (see II(A), Art. 8).

(86)
$$u_t + uu_x + \frac{c^2}{\rho}\rho_x = 0$$
$$\rho_t + \rho u_x + u\rho_x = 0$$

We shall omit further details concerning the interesting theorotical mathematical aspects of the topic. It may be emphasized, however, that the numerical solution of our initial value problem offers little difficulty since methods of finite differences may be effectively employed.

III. Appendix 1

DETONATION WAVES

50. Detonation conditions. Closely related to shock waves are detonation waves*, i.e., waves in an explosion process which consist of fast-moving discontinuity surfaces separating the unexploded part of the explosive from the explosion products. The fact that the equation of state of the medium is different on the two sides of the front does not constitute an essential difference between these waves and the shock waves considered in Chapter III. Obviously the two mechanical discontinuity conditions (Art. 32) which express the laws of conservation of mass and momentum subsist for detonation fronts as well. An essential modification, however, arises in the third, the thermodynamic condition, inasmuch as now the law of conservation of energy requires an additional term in the energy balance, representing the energy set free by the chemical reaction. This liberated chemical energy per unit mass is a quantity f characteristic of the explosive material.

Let us indicate the unexploded substance by the symbol (o), the burnt gases by the symbol (1). Then, in the notation of Art. 28, Chapter III, the thermodynamic discontinuity condition becomes

(87) $$\frac{1}{2} v_o^2 + i_o = \frac{1}{2} v_1^2 + i_1 + f .$$

Using the mechanical shock conditions to eliminate v_o^2 and v_1^2 (see derivation of III(iii_{***}), Art. 28), this can be written

(88) $$\frac{1}{2}(p_1 - p_o)(\tau_1 + \tau_o) - i_1 + i_o = f.$$

---

* For a discussion of the detonation process particular reference is made to the paper by Becker [18]. See also the report by von Neumann [19], and the literature quoted there.

The two mechanical conditions and condition (87) or (88) will be called the detonation conditions.

51. Auxiliary hypothesis. We recall that an ordinary shock wave is determined by the state on the front side (o) and one additional quantity referring to the situation on the back side (1), e.g., the speed of a piston. To maintain a shock wave some externally acting agent, e.g., a piston, is necessary as a source of energy. With a detonation wave, however, the physical situation is different. The detonation process maintains itself by means of the liberated chemical energy. While the whole process in a shock wave between a front state (o) and a back state (1), for example, is determined by the knowledge of the state (o) and the "piston speed" u_1 behind the wave, a detonation wave racing into the explosive in the state of rest (o) cannot be determined in the same way since nothing about the state (1) behind the shock front is known in advance. Hence a further condition is needed for characterizing the detonation wave. Such additional conditions were proposed by Chapman (1899) and independently by Jouguet (1905), the two proposals later being recognized as equivalent. A completely convincing rational derivation of the Chapman-Jouguet hypothesis has apparently not yet been given, but the hypothesis can be made plausible and agrees well with experiments.

The detonation condition together with a given state (o) characterizes a one-parametric family of possible discontinuity transitions from state (o) to a virtual* state (1). Now the Chapman-Jouguet hypothesis singles out from this family of virtual transitions two particular cases as describing the actual detonation process.

The hypothesis will be formulated in three different equivalent forms A, B, C.

(A) Jouguet's hypothesis. The flow of the exploded gases immediately behind the front is exactly sonic, i.e.,

---

* Called "virtual" because, of the one-parametric manifold of states (1) so determined, only two are actually possible.

(89) $$v_1 = c_1 .$$

We shall presently see why this is a plausible condition. At the moment we conclude from $v_1^2 = \tau_1^2 m^2 = -\tau_1^2 \frac{p_1 - p_0}{\tau_1 - \tau_0}$ (see Art. 32) that (89) is equivalent to

(90) $$\frac{p_1 - p_0}{\tau_1 - \tau_0} = -\rho_1^2 c_1^2 .$$

(B) Chapman's hypothesis. The velocity $-v_0$ of the detonation front relative to the undetonated explosive ahead of it is an extremum (maximum or minimum) as compared with the velocities of virtual discontinuity fronts relative to the explosive. Then from the first condition, $\rho_0 v_0 = \rho_1 v_1 = m$, we have

$$\rho_0^2 v_0^2 = m^2 = -\frac{p_1 - p_0}{\tau_1 - \tau_0} .$$

Hence property B is equivalent to the property that the ratio $\frac{p_1 - p_0}{\tau_1 - \tau_0}$ be an extremum, p_1 and τ_1 being coupled by relation (88). Accordingly, the analytical expression of property B is

(91) $$\frac{dp_1}{d\tau_1} = \frac{p_1 - p_0}{\tau_1 - \tau_0} .$$

Geometrically speaking, the straight line connecting the point (τ_0, p_0) with the point (τ_1, p_1) in the τ, p-plane is tangential to the graph of the relation (88) for $\tau = \tau_1$, $p = p_1$.

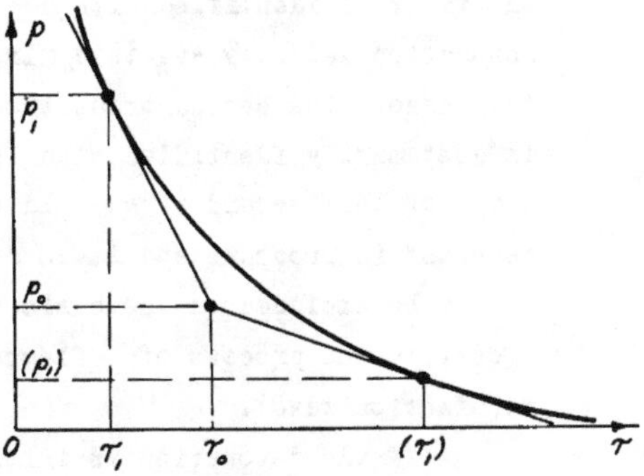

Figure 60
Geometrical representation of Chapman's hypothesis (B).

(C) A third formulation is the following. The entropy of the state (1) behind the detonation front is an extremum (maximum or minimum) when compared with all virtual states that satisfy the three detonation conditions, the state (o) being held fixed. The analytical formulation of this is

$$\frac{dp_1}{d\tau_1} = -\rho_1^2 c_1^2 , \tag{92}$$

since $c^2 = \frac{dp}{d\rho} = -\tau^2 \frac{dp}{d\tau}$ when the differential of the entropy vanishes.

To show that property C entails A and B we note that $\frac{di_1}{d\tau_1} = \tau_1 \frac{dp_1}{d\tau_1} = -\rho_1 c_1^2$ when the differential of the entropy vanishes. Hence on differentiating relation (88) we obtain the relation

$$\frac{1}{2}(\tau_1 + \tau_0)\frac{dp_1}{d\tau_1} + \frac{1}{2}(p_1 - p_0) - \tau_1 \frac{dp_1}{d\tau_1} = 0,$$

which is equivalent to (91) and, by virtue of $\frac{dp_1}{d\tau_1} = -\rho_1^2 c_1^2$, to (90).

That A or B entails the two remaining conditions is easily shown in a similar manner.

Of the two possible transitions characterized by these hypotheses the one for which pressure and density are increased beyond the front is identified with a <u>detonation</u> in the proper sense. The detonation velocity $-v_0$ is a minimum and the entropy a maximum in this case. The second transition leading to lower pressure and density is customarily identified with the process of <u>deflagration</u>.* Transitions of this second type would correspond to expansive shocks with decrease in pressure and it is doubtful whether or not such transitions should be excluded on the basis of the second law of thermodynamics. (Possibly the process of deflagration should rather be described as a rarefaction wave).

If the detonation is initiated by the motion of a piston which

---

* Von Neumann [19] has advanced the idea that a detonation consists rather in a non-reactive shock followed by a deflagration process, with the same result that a single detonation shock would give.

comes to rest after a short time, a rarefaction wave follows the detonation wave through which the burnt gases are decelerated until they come to rest. A rarefaction wave following a non-reactive shock wave overtakes and devours it (see Art. 45). If a rarefaction wave follows a detonation wave, however, its head may stay in contact with the detonation front without interference, since both the head of the rarefaction wave and the detonation front move with the same velocity relative to the gas behind the front. This is a very significant consequence of Jouguet's hypothesis A and makes it plausible.

The rarefaction wave is here assumed to be one-dimensional as if the burnt gases were confined in a cylindrical tube. The walls of the tube must be rather strong to withstand the high pressure in the burnt gases. In frequent cases, however, the walls will yield and offer negligible resistance. Then, observed from the detonation front, the burnt gases flow as if they came out of an orifice.

III. Appendix 2

WAVE PROPAGATION IN ELASTIC-PLASTIC MATERIAL*

52. The medium. Solid matter is capable of elastic deformations under certain conditions, and of plastic changes in shape under others. The property of matter characterizing it as elastic or plastic can be expressed mathematically by the relation existing between the stress and the strain, and will be defined in the following paragraphs (see also Art. 2(b), Chapter I).

In such elastic-plastic materials an important variant of wave propagation occurs which differs in many respects from wave motion in gases. The decisive new feature is that shock waves and continuous simple waves occur in both expansive and compressive motion. It is also interesting that there is always a sonic discontinuity at the head of a rarefaction wave entering a zone in which the material is unstrained. In contrast to gas, which expands indefinitely under zero pressure, an elastic-plastic material assumes a well-defined original state when it suffers no stress.

The Lagrangean representation seems the natural one to employ for the treatment of motion in such material. Let us consider an elastic-plastic cylindrical bar of uniform cross-section in its original (unstrained) state. When the bar is deformed in the direction of the axis, the axial coordinate x of a particle depends on its "original" abscissa a and on the time t: $x = x(a,t)$. The strain is then defined in terms of the rate of change $x_a = \frac{\partial x}{\partial a}$ by

$$\varepsilon = x_a - 1 \; . \tag{93}$$

When ρ is the density of mass and ρ_0 the "original" density, we clearly have $\rho_0 da = \rho dx$ or

* For the theory of elastic-plastic waves, see a great number of reports and memoranda by v. Karman and others issued through Division 2 of the NDRO.

$$\frac{\rho_0}{\rho} = (1 + \varepsilon). \tag{94}$$

The stress is the force per unit area acting in normal direction against a cross-section; for the following considerations, however, a somewhat different quantity is to be used. Actually, the motion of the bar does not take place solely in the axial direction since an extension in the axial direction is always connected with a contraction in the perpendicular direction. Thus for the desired approximate one-dimensional treatment, the significant quantity is not the stress, but the total force acting in the normal direction against a cross-section. This total force divided by the constant area of the original cross-section, the so-called engineering stress, is the one denoted in the following by σ and simply called the stress. This stress is then assumed to be a known function of the strain

$$\sigma = \sigma(\varepsilon), \tag{95}$$

this function depending only on the nature of the material. One always has

$$\sigma \gtreqless 0 \quad \text{for} \quad \varepsilon \gtreqless 0, \tag{96}$$

that is, the stress is positive in tension and negative in compression ($\sigma = 0$ for $\varepsilon = 0$ is true by definition). For most materials the further inequality

$$\frac{d\sigma}{d\varepsilon} > 0 \tag{97}$$

is satisfied throughout; that is, increasing strain implies increasing stress. In the following discussion we assume relation (97) to hold.

A material is called elastic when the stress depends linearly on the strain. Most materials are elastic when the strain does not exceed a certain limit, the critical strain ε_*. The stress-strain relation is then

(98) $$\sigma = E\varepsilon, \qquad |\varepsilon| \leq \varepsilon_*,$$

the constant E being Young's modulus.

A material is here called plastic when the stress is a non-linear function of the strain, the latter being greater than the critical strain. For the plastic region we assume

(99) $$0 < \frac{d\sigma}{d\varepsilon} < E, \qquad |\varepsilon| > \varepsilon_*$$

and

(100) $$\frac{d^2\sigma}{d\varepsilon^2} \begin{cases} < 0 & \text{for } \varepsilon > \varepsilon_* \\ > 0 & \text{for } \varepsilon < -\varepsilon_*. \end{cases}$$

It should be noted that for some materials there is a certain range of values of the strain where the stress is independent of the strain but depends on the rate of strain. Some authors reserve the notion "plastic" for such a state of the material. We have excluded these cases by condition (97). A typical function $\sigma = \sigma(\varepsilon)$ is indicated in the accompanying graph (Fig. 61a).

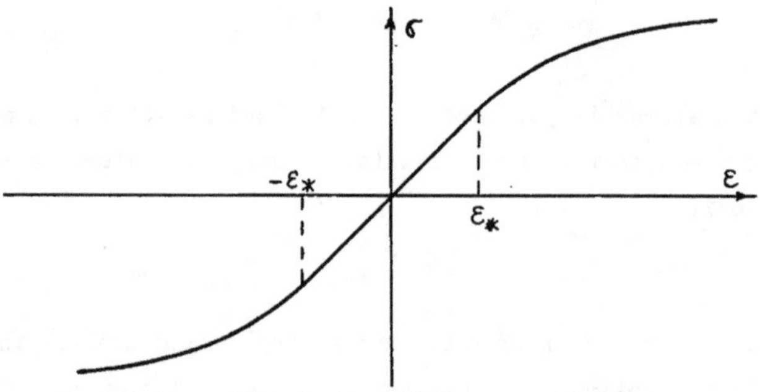

Figure 61a
Graph of stress-strain relationship for elastic-plastic material.

It is interesting to compare the stress-strain relation for an elastic-plastic material with the adiabatic pressure-density relation for a polytropic gas. To this end we identify the pressure with the negative stress, $p = -\sigma$ (although this is not quite proper since σ is the "engineering stress"). We further set $\rho = \frac{\rho_0}{(1 + \varepsilon)}$ in accordance with (94) Then the adiabatic relation for a gas becomes

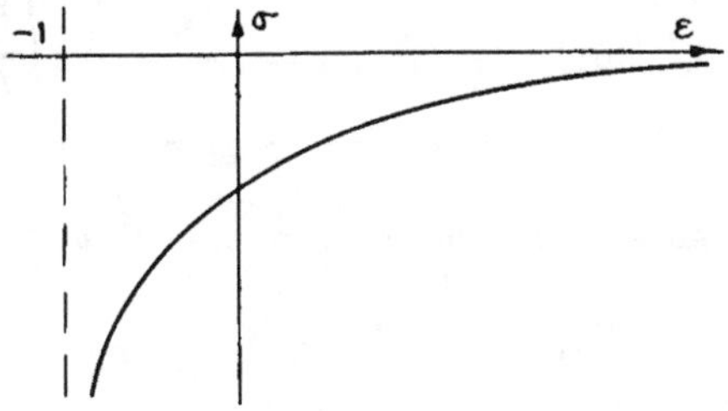

Figure 61b
Graph of "stress-strain" relationship for a gas ($\sigma = -p$).

$$\sigma = - \frac{p_0}{(1 + \varepsilon)^{\gamma}},$$

the graph of which is given in Fig. 61b . We observe that for tensile strain, $\varepsilon > 0$, the trend of the two ε,σ-curves is the same, in that $\frac{d\sigma}{d\varepsilon}$ decreases for increasing ε. However, for compressive strain, $\varepsilon < 0$, the slope $\frac{d\sigma}{d\varepsilon}$ decreases as ε decreases for elastic-plastic material while for gas it increases as ε decreases. The significance of this fact will become apparent in the following articles.

53 . The equation of motion. The motion of a particle with the original abscissa a is given by a function $x = x(a,t)$; its velocity is therefore given by

(101) $$u = x_t = \frac{\partial x}{\partial t}.$$

The equation of motion, $\rho u_t = \frac{\sigma_a}{x_a}$, becomes by (93) and (94)

(102) $$u_t = g^2 \varepsilon_a$$

with

(103)
$$g = \sqrt{\frac{\frac{d\sigma}{d\varepsilon}}{\rho_0}}.$$

As a second equation we have from (93) and (101)

(104)
$$\varepsilon_t = u_a.$$

The difference between the present equations and the form of the Lagrange equations which we have employed for gases (see Art. 7, Chapter I) is that ε is used instead of $\tau = \frac{1}{\rho}$ and a instead of $h = \rho_0 a$.

The quantity g is clearly the rate of change $\frac{da}{dt}$ with which a disturbance shifts from particle to particle. We call a rate of change $\frac{da}{dt}$ a <u>shift rate</u> and $g(\varepsilon)$ is in particular called the <u>characteristic shift rate</u>. The shift rate g is connected with the sound speed c and the impedance $k = \rho c$, previously defined for gases (see Art. 7, Chapter I), by the relations

(105)
$$g = \frac{k}{\rho_0} = \frac{\rho}{\rho_0} c,$$

where c is defined by

(106)
$$c = \sqrt{-\frac{d\sigma}{d\rho}}.$$

In the elastic range, the shift rate

(107)
$$g_0 = \sqrt{\frac{E}{\rho_0}}$$

is constant (while the sound speed $c = \frac{\rho_0}{\rho} g_0$ is not). The graph of the characteristic shift rate $g(\varepsilon)$ is given below. In accordance with assumption (100), $g(\varepsilon)$ decreases during tension when ε becomes larger than ε_*, and also during compression when ε becomes smaller than $-\varepsilon_*$.

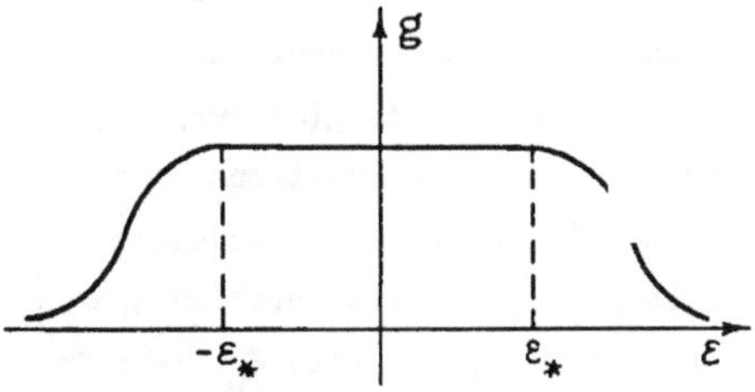

Figure 62a
Graph of relationship between shift rate g and strain ε for elastic-plastic material.

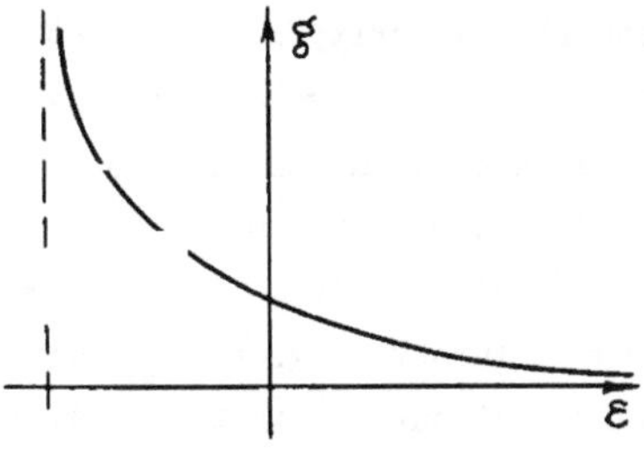

Figure 62b
Graph of relationship between "shift rate" and "strain" for a gas.

54 . Impact loading. The basic problem of wave propagation in a bar of elastic-plastic material is concerned with the motion resulting from impact loading, i.e., from a velocity being suddenly imparted to one end of the bar and then maintained there. Pushing in or pulling out the end of the bar corresponds to pushing in or withdrawing the piston in a tube filled with gas. From a receding piston a centered rarefaction wave is propagated into the gas. We proceed to discuss the corresponding phenomenon when the end of the bar is pulled out.

Imparting a constant velocity u_0 to the end cross-section is, as we shall see, equivalent to imparting to it a constant strain ε_0. If this strain ε_0 is less than the critical strain ε_*, the strain resulting in the bar through wave propagation also remains below the critical strain. The wave propagation is, therefore, governed by linear differential equations with constant coefficients. If such is the case, as was explained earlier (Art. 11, Chapter II), initial discontinuities are propagated as discontinuities with constant characteristic shift rate.

Whenever the initial strain ε_0 is greater than the critical strain, the differential equations of propagation are non-linear, and non-linearity implies that whether the initial discontinuity is propagated through a shock wave or smoothed out by a rarefaction wave depends on whether the characteristic shift rate $g(\varepsilon)$ increases or decreases with increasing ε. According to the assumptions made here (see (100)), $g(\varepsilon)$ decreases when $\varepsilon > \varepsilon_*$ increases. Consequently the influence of greater values of ε is propagated with smaller speed. This fact entails that a suddenly imparted initial strain ε_0, when it is greater than the critical strain, is propagated through a rarefaction wave.

In order to determine the resulting motion it is convenient to write the equations of motion in characteristic form (see Art. 9, Chapter II),

$$da = \pm g dt, \tag{108}$$

$$du \mp g d\varepsilon = 0. \tag{109}$$

Introducing the function

$$\phi(\varepsilon) = \int_0^{\varepsilon} g(\varepsilon) d\varepsilon, \tag{110}$$

we can write equation (109) in the form

$$d(u \mp \phi) = 0. \tag{111}$$

Suppose now that the bar lies along the positive x-axis, $x \geq 0$, and the impact produces a velocity $u_0 < 0$ at the end cross-section $x = 0$. A centered simple wave will move in forward direction. Across it $u + \phi(\varepsilon)$ is constant; since $u = 0$, $\phi(\varepsilon) = 0$ for $t = 0$, $x > 0$, we have, across the wave,

$$u + \phi(\varepsilon) = 0. \tag{112}$$

In particular, the strain ε_0 produced by the impact at the end cross-section is such that

(113) $$u_0 = -\phi(\varepsilon_0).$$

The quantity $\phi(\varepsilon)$ is called the impact velocity because $-\phi(\varepsilon)$ is the velocity u that must be imparted to the end of the bar in order to produce the strain ε there.

The influence of the impact travels with the shift rate g_0; hence we have

(114) $$\varepsilon = 0, \quad u = 0 \quad \text{for} \quad 0 \leq t \leq \frac{a}{g_0}.$$

If $\varepsilon_0 \leq \varepsilon_*$, the strain (of a particle with original position a) jumps from $\varepsilon = 0$ to $\varepsilon = \varepsilon_0$ at the time $t = \frac{a}{g_0}$, while the velocity jumps from $u = 0$ to $u = -\phi(\varepsilon_0)$. Afterwards the state remains constant;

(115) $$\varepsilon = \varepsilon_0, \quad u = -\phi(\varepsilon_0) \quad \text{for} \quad t \geq \frac{a}{g_0}.$$

If, however, $\varepsilon_0 > \varepsilon_*$, the strain jumps from $\varepsilon = 0$ to $\varepsilon = \varepsilon_*$ at the time $t = \frac{a}{g_0}$, while the velocity jumps from $u = 0$ to $u = -\phi(\varepsilon_*)$. Afterwards there is a simple centered rarefaction wave which can be described by the parametric representation

(116) $$u = -\phi(\varepsilon), \quad \frac{a}{t} = g(\varepsilon), \quad \text{for} \quad \varepsilon_* < \varepsilon \leq \varepsilon_0.$$

From this representation to every time in the interval $\frac{a}{g(\varepsilon_*)} < t < \frac{a}{g(\varepsilon_0)}$ values ε and u are uniquely assigned, since $g(\varepsilon)$ decreases with increasing ε according to assumption (100). After the impact strain $\varepsilon_0 = -\phi(u_0)$ has been reached, the state remains constant; in other words

(117) $$\varepsilon = \varepsilon_0, \quad u = -\phi(\varepsilon_0) \quad \text{for} \quad t > \frac{a}{g(\varepsilon_0)}.$$

The motion would, of course, also be described by formulating the occurrences at a fixed time $t = t_1$. Fig.63b corresponds to such a description.

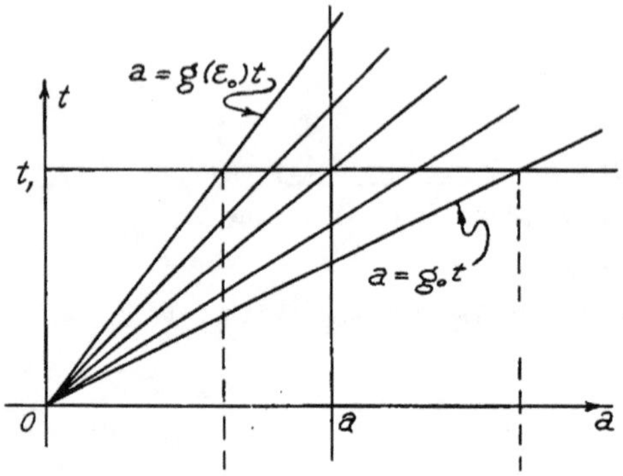

Figure 63a
Centered simple wave in elastic-plastic material.

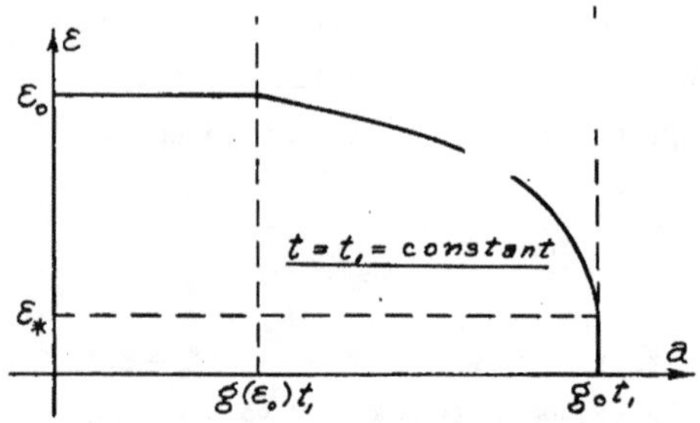

Figure 63b
Distribution of strain in elastic-plastic material at time t_1 after tensile impact.

The discontinuity at the head of the wave is of particular interest. It does not deserve the name "shock" since it is "sonic", moving with the characteristic shift rate. It may be considered as a degenerate section of a simple wave due to the fact that all the characteristics $a = g(\varepsilon)t$ for $0 \leq \varepsilon \leq \varepsilon_*$ have the same slope and thus coincide.

We now discuss the case of initial impact producing a <u>positive</u> velocity u_0 and hence a compressive strain $\varepsilon_0 < 0$. This case corresponds to a piston being suddenly pushed into a gas-filled tube. In the gas, as we know, a shock wave results. In the elastic-plastic material, however, the compression is propagated through a simple wave with a discontinuous head just as a tensile impact would be propagated. The description of the wave can be obtained from that of the expansion wave by substituting -u for u in formulas (112) to (117). That the description of the simple wave derived from (116) is such that to every time in the interval $\frac{a}{g(-\varepsilon_*)} < t \leq \frac{a}{g(\varepsilon_0)}$ values of ε and u are uniquely assigned, follows again from the fact that $g(\varepsilon)$ decreases as ε decreases from $-\varepsilon_*$ to ε_0 according to assumption (100).

Quite generally, an initial discontinuity, as stated before, is propagated through a simple wave if, and only if, the characteristic shift rate g for the state ahead of the discontinuity is greater than the shift rate for the state behind the discontinuity. For the material considered we saw that this is the case for both a tensile and a compressive impact since $g(\varepsilon)$ decreases as $|\varepsilon|$ increases from ε_* on. In a gas, however, $g(\varepsilon)$ increases with decreasing ε; therefore a compressive impact in a gas is propagated through a shock wave.

It may be mentioned that there are materials for which assumption (99) is not satisfied and $\frac{dg}{d\varepsilon}$ changes sign for sufficiently large strains. Then, if the impact is strong enough, the transition is propagated by a simple wave followed by a shock, the state in front of the shock being so determined that the characteristic shift rate of this state coincides with the shift rate of the shock wave.

55. Stopping shocks There is another peculiar situation in which shocks are propagated through an elastic-plastic material. So far, it has been assumed that the velocity imparted to one end is maintained there indefinitely by applying the appropriate stress, $\sigma = \sigma(\varepsilon_0)$. It is, of course, important to investigate what happens when this stress is suddenly released. The influence of this new discontinuity can certainly not be propagated through a simple wave since the characteristic shift rate g is smaller before stopping than after stopping. It is therefore to be expected that the influence of this stopping is propagated through a shock wave. This shock is of a particularly simple type due to the following circumstances connected

with the phenomenon of <u>hysteresis</u>. When a plastic material is released from a strained position it will not, on its return, obey the same stress-strain relation as when the strain was produced. The general experience is that on returning, the stress depends linearly on the strain and that $\frac{d\sigma}{d\varepsilon} = E$ as in the elastic state (see Fig. 64). Therefore, when the stress has returned to zero, a "permanent" strain different from zero remains. The same then is true for the transition across a stopping shock. Let $[\sigma]$ and $[\varepsilon]$ be the differences of the values of σ and ε, respectively, in front of and in back of the shock. Then, according to the property $\frac{d\sigma}{d\varepsilon} = E$ formulated above,

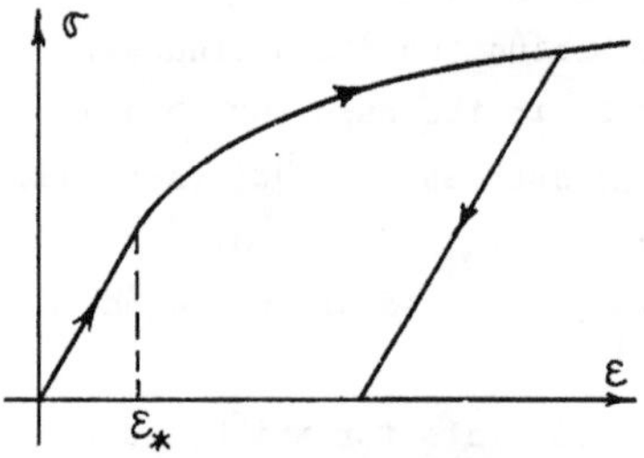

Figure 64
Stress-strain graph illustrating phenomenon of hysteresis in plastic material.

$$(118) \qquad [\sigma] = E\,[\varepsilon] .$$

The shock transition relations for the Lagrangean representation were derived in Art. 40, Chapter III. The first two of them can be written

$$(119) \qquad [u] = -\dot{a}[\varepsilon], \qquad [\sigma] = -\rho_0\dot{a}[u],$$

where $\dot{a}$ is the shift rate of the shock, $\rho_0\dot{a}$ is the mass flux per unit area crossing the shock from front to back. Eliminating $[u]$ we find

$$(120) \qquad [\sigma] = \rho_0\dot{a}^2[\varepsilon] .$$

Hence $\dot{a}^2 = \frac{E}{\rho_0}$ or $\dot{a} = g_0$. Thus it is seen that the shift rate $\dot{a}$ of the shock coincides with the characteristic shift rate g_0 belonging to the elastic state.

The third shock relation, expressing conservation of energy, can now be used to determine energy changes; but the shock is already determined by the first two conditions alone. It is in this respect that the present "stopping shock" is simpler in character than the shocks occurring in gas dynamics.

The decisive feature of gas dynamical shocks is that they produce permanent changes in the conditions of the gas by increasing the entropy. One is tempted to consider the change of entropy as analogous to the permanent strain resulting after a stopping process. This analogy, however, does not carry very far. Permanent changes in elastic-plastic material appear to be linked with the non-linear phase of the process; in contrast to gases, they would also occur if the stress were reduced in a gradual manner. Therefore, permanent deformations can not be ascribed to the shock transition as such.

56. Interactions and reflections. The stopping shock eventually catches up with the simple wave running ahead and a more complex process of interactions will ensue. Due to the simple nature of the shock it is possible to analyse this process of interaction in all detail. This has been done, but we shall refrain from reporting on the results here, mentioning only that the final permanent change of state of the material can be determined completely.

Wave motion in elastic-plastic material has also been analyzed in another direction. The motion in a bar of finite length can be described by a succession of reflections. It is appropriate to introduce as new variables the velocity u and the impact velocity $\phi = \phi(\varepsilon)$. Then equations (102) and (104) go over into the linear equations

$$(121) \qquad a_{\phi} = g t_{u}, \qquad a_{u} = g t_{\phi} ,$$

where g may be considered a function of ϕ. When the other end of the bar, $x = \ell$, is fixed, the velocity there is $u = 0$; hence the region in the u,ϕ-plane is the fixed strip

$$u_0 \leq u \leq 0, \qquad 0 \leq \phi .$$

It is to be borne in mind that in the u,ϕ-plane the image of a constant state is a point and the image of a simple wave is a line. The image of a region of interaction between incoming and reflected waves is a triangle. The motion corresponding to this triangle can then be determined by an approximate method using characteristic lines.

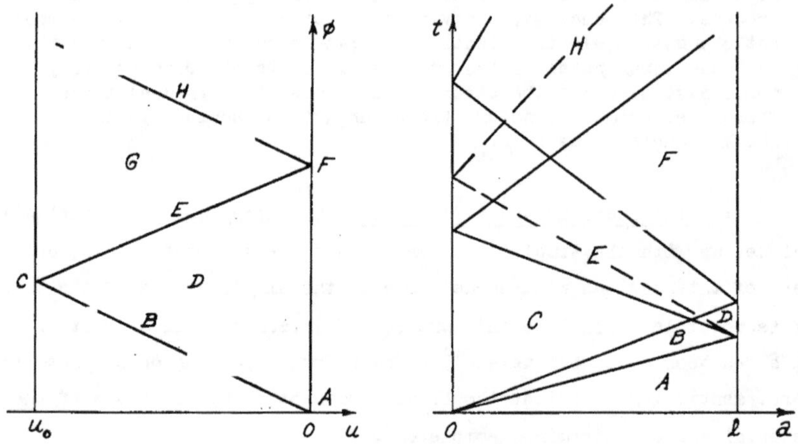

Figure 65
Graphical representation of motion of a bar as a succession of reflected waves in u,ϕ-plane and a,t-plane.

It is seen that on successive reflections the strain increases. Accordingly, the characteristics which coincided in the first simple wave and formed the elastic discontinuity front will, when continued through reflection, come into the non-linear range and thus spread. Therefore the reflected waves have a continuous front. The characteristic line resulting through reflection from the one with $\varepsilon = \varepsilon_*$ is shown as a dotted line in Fig.65.

One more remark might be made in conclusion. The nature of the stress-strain relation is in reality not so well established as that for gases and varies considerably for different materials.

Various approximate assumptions can be made. In particular, the relation $\sigma = \sigma(\varepsilon)$ can with sufficient approximation be so chosen that an explicit integration of the differential equation becomes possible. The assumption

$$g = \frac{b^2 g_0}{(b + \varepsilon)^2} ,$$

for instance, has proved to be very suitable for this purpose. In particular, the problem of reflection can be treated rather explicitly under this assumption and the final state approached as time increases indefinitely can be determined.

III. Appendix 3

WAVE MOTION IN OPEN CHANNELS

57. Another example of non-linear wave motion is encountered in the motion of an incompressible fluid in an open channel. The channel is said to be open if the top surface of the fluid is free and the pressure there equals atmospheric pressure. When the height of the fluid is small compared with the lateral extension of the channel, the problem may be simplified by considering only average values. It will be sufficient to consider the one-dimensional case.

One then considers the fluid as moving in the x,z-plane, the bottom of the channel being $z = 0$, the top surface being given by $z = h(x,t)$. Denoting the components of velocity in the x and z directions by $u(x,z,t)$ and $w(x,z,t)$ respectively, the continuity equation $u_x + w_z = 0$ holds in the interior, while at the top surface one has $h_t + uh_x = w$. Integrating the former equation from $z = 0$ to $z = h$ and using the latter equation at $z = h$, one finds

$$h_t + \frac{\partial}{\partial x}\int_0^h u\,dz = h_t + uh_x + \int_0^h u_x\,dz = 0. \tag{122}$$

When one now introduces the average velocity

$$\bar{u} = \frac{1}{h}\int_0^h u\,dz \tag{123}$$

and the mass per unit area

$$P = \rho g h \tag{124}$$

(ρ being the mass per unit volume, g the acceleration due to gravity), the average continuity equation can be written

(125) $$P_t + (P\bar{u})_x = 0.$$

Let p be the excess pressure over that of the atmosphere,

(126) $$\mathcal{P} = -\int_0^h p dz$$

the excess force per unit length. Let a be the x-component of the acceleration,

(127) $$\bar{a} = \frac{1}{h}\int_0^h a dz,$$

the average acceleration. Then, by integrating the relation $\rho a g = -p_x$, we obtain

(128) $$P\bar{a} = \mathcal{P}_x$$

where $p = 0$ at the top surface has been used. The equations (125) and (128) are so far exact. An approximation is introduced when the relation

(129) $$\bar{a} = \bar{u}_t + \bar{u}\bar{u}_x$$

is assumed, which, of course, does not result by integration from $a = u_t + uu_x + wu_z$.

Relations (125), (128), (129) are the same as those for compressible fluids, P and $\mathcal{P}$ taking the place of density and pressure To complete the system we need a relation between P and $\mathcal{P}$. Assuming that the influence of the z-component of the acceleration on the pressure is small, we may assume that the pressure increases linearly with the distance from the top surface

(130) $$p = g\rho(h - z),$$

g being the acceleration of gravity. By integration we then find

(131) $$\mathcal{P} = \frac{g}{2}\rho h^2 = \frac{P^2}{2\rho g}.$$

Thus the relation between P and $\mathcal{P}$ is of the form $\mathcal{P} = cP^\gamma$, with $\gamma = 2$. Thus continuous motion in open channels corresponds closely to that in gases.* If the fluid is retracted at one end of the channel, then the water sinks to a lower level and this depression is propagated through a "rarefaction" wave into the channel. If the fluid is pressed into the channel, a shock wave producing a sudden increase in velocity and altitude travels into the channel. Since the density-pressure relation does not depend on an additional parameter, such as the entropy in gases, the shock transition is completely determined by the mechanical shock conditions.**

It is clear that the analogy to two-dimensional gas motion could be carried through for two-dimensional flow in open channels as well.

---

* The analogy with gas flow was first mentioned by Jouguet and worked out by Riaboushinsky (see v. Kármán [53]).

** In water-like substances (liquids), the third shock condition could be used for determining the heat loss through the shock necessary to restore the energy balance. For restoring the energy balance in the present case it would be necessary to refine the approximation process used so far, which consists of taking simple averages.

IV. ISENTROPIC IRROTATIONAL STEADY PLANE FLOW. OBLIQUE SHOCK FRONTS. SHOCK REFLECTION.

58. Introduction. Next in simplicity to the theory of one-dimensional flow is that of flow in a plane or two-dimensional or plane flow, provided that the flow is steady, irrotational (Art. 5, Chapter I) and isentropic (Art. 4). Many important phenomena can be understood, at least qualitatively, on the basis of a theory of two-dimensional steady flow. In this chapter the mathematical theory of such a flow will be developed along lines similar to those followed in Chapter III in the case of one dimension and similar completeness can thus be achieved.

Let us recall the analytical background. Under our assumptions the flow is characterized by the two components u,v of the flow velocity $\vec{q}$ as functions of the Cartesian coordinates x,y in the plane; likewise, ρ, p, c are functions of x,y alone, not depending on z or on t. There exists for the flow a constant c_*, the critical speed, so that Bernoulli's equation

$$(1)\qquad \mu^2(u^2 + v^2) + (1 - \mu^2)c^2 = c_*^2 = \mu^2\hat{q}^2,$$

with $\mu^2 = \dfrac{\gamma - 1}{\gamma + 1}$, holds throughout provided that we assume the medium to be a polytropic gas as we shall do in this chapter. The differential equations can be written (see I(35), Art. 6) in the form

$$(2)\qquad u_y - v_x = 0$$

$$(3)\qquad (\rho u)_x + (\rho v)y = 0.$$

By eliminating ρ (see Art. 6, Chapter I) equation (3) becomes

$$(3')\qquad (c^2 - u^2)u_x - uv(u_y + v_x) + (c^2 - v^2)v_y = 0$$

which together with (2) and (1) formulates the general laws for the flow, assuming the critical speed c_* as a parameter given from the outset.

By introducing a velocity potential $\phi(x,y)$ with $\phi_x = u$ and $\phi_y = v$, equation (3') becomes a differential equation of second order for ϕ,

$$(4) \qquad (c^2 - \phi_x^2)\phi_{xx} - 2\phi_x\phi_y\phi_{xy} + (c^2 - \phi_y^2)\phi_{yy} = 0$$

which must be considered together with (1).

As pointed out in Art. 8, Chapter II, the differential equations (2) and (3') are transformed into a system of linear differential equations upon introducing u,v as independent, x,y as dependent variables, provided that for the solution considered the Jacobian $u_x v_y - u_y v_x$ does not vanish.*

The transformation of the differential equations (2), (3') by the introduction of characteristic parameters α,β as developed in the general theory of Art. 9, Chapter II, leads to the following characteristic forms by a straightforward substitution of the coefficients of (2), (3) into the formulas of Art. 9:

$$I_+ \quad y_\alpha = \zeta_+ x_\alpha \qquad I_- \quad y_\beta = \zeta_- x_\beta$$

$$II_+ \quad u_\alpha = -\zeta_- v_\alpha \qquad II_- \quad u_\beta = -\zeta_+ v_\beta$$

---

* For the differential equation of second order (4) this change of variables becomes a <u>Legendre transformation</u>. By virtue of $x_v = y_u$ we have a new potential $\chi(u,v)$ with $\chi_u = x$ and $\chi_v = y$, so that

$$\chi = \int(x\,du + y\,dv) = xu + yv - \int(u\,dx + v\,dy) = xu + vy - \phi .$$ Thus

the linearization of the differential equation of second order is effected by the Legendre transformation $\phi_x = u$, $\phi_y = v$, $\chi = xu + yv - \phi$, which is inverted by $\chi_u = x$, $\chi_v = y$.

where $\zeta_+ = -\frac{uv + cw}{c^2 - u^2}$, $\zeta_- = -\frac{uv - cw}{c^2 - u^2}$ $(w^2 = u^2 + v^2 - c^2)$, c^2 being given in terms of u and v by Bernoulli's law.

In the notation of Chapter II, β = constant gives the characteristics C_+ in the x,y-plane and Γ_+ in the u,v-plane (hodograph plane), while α = constant gives the characteristics C_- and Γ_- in the respective planes. The relations

$$(5) \qquad x_\alpha u_\beta + y_\alpha v_\beta = 0, \qquad x_\beta u_\alpha + y_\beta v_\alpha = 0$$

expressing orthogonality of C_+ on Γ_- and of C_- on Γ_+ follow immediately.

Equations (II) determine the families of characteristics Γ_+ and Γ_- irrespective of the specific flow as the fixed curves satisfying the ordinary differential equations $\frac{du}{dv} = -\zeta_-(u,v)$ and $\frac{du}{dv} = -\zeta_+(u,v)$ respectively.*

As remarked in Chapter II, the transformation to characteristic parameters is possible in this case only if

$$u^2 + v^2 = q^2 > c^2,$$

i.e., only for <u>supersonic flow</u>. Unless otherwise stated, we shall henceforth make this assumption. Only then can the solutions of the differential equations be interpreted in terms of "wave propagation", implying the occurrence of definite ranges of influence and domains of dependence.

A few remarks, though partly repititious, may be inserted here. From Bernoulli's law it is clear that the relations between the critical

---

* Having obtained an integral of these ordinary differential equations, one can integrate the system of the two partial differential equations (II) completely by expressions involving two arbitrary functions; substitution in (I) then leaves two linear partial differential equations of first order whose solution is possible by simple iteration (see Art. 10, Chapter II). We are, however, more interested in solving specific problems than in a "general" solution of the differential equations.

speed c_*, the limit speed $\hat{q}$ for the flow and the greatest value c_{max} which the sound speed in the flow might conceivably attain (in a state of stagnation, in which the flow speed $q = 0$) are

$$(6) \qquad c_{max} = c_* \sqrt{\frac{\gamma + 1}{2}} = \hat{q}\sqrt{\frac{\gamma - 1}{2}} \; .$$

To the critical speed c_* there corresponds a critical pressure p_* and a critical density ρ_* connected by the relations

$$(7) \qquad c_*^2 = \gamma \frac{p_*}{\rho_*} \; , \qquad p_* = A\rho_*^{\gamma} \; .$$

There also exist for a given c_* a stagnation pressure p_{max} and a stagnation density ρ_{max}, which are the maxima of pressure and density, also attained in a state of stagnation ($q = 0$) of the flow, and our relations imply the following relations between these quantities:

$$(8) \qquad \frac{p}{p_*} = \left(\frac{\rho}{\rho_*}\right)^{\gamma} = \left(\frac{c}{c_*}\right)^{\frac{2\gamma}{\gamma - 1}}$$

$$(9) \qquad \frac{p_{max}}{p_*} = \left(\frac{c_{max.}}{c_*}\right)^{\frac{2\gamma}{\gamma - 1}} = \left(\frac{\gamma + 1}{2}\right)^{\frac{\gamma}{\gamma - 1}} \qquad (= 1.893 \text{ for air})$$

where (6) and (7) have been employed.

It is interesting that our relationships determine rather narrow margins within which the quotients $\frac{c}{c_*}$, $\frac{p}{p_*}$ and $\frac{\rho}{\rho_*}$ may vary in an isentropic flow, this margin being determined solely by the adiabatic exponent γ. Thus, the sound speed in air is always less than 10% above the critical speed for a given flow (the latter, of course, may be very large relative to the sound speed in air at rest and at atmospheric pressure and normal temperature).

Isentropic steady flow is reversible, i.e., to any such flow there corresponds another flow in which the pressure, density and temperature remain unchanged while the direction of the flow velocity is reversed.

Finally, we should bear in mind that the steady character of a flow must be understood with respect to a definite coordinate system. Observed from a coordinate system moving at constant velocity relative to the original one, the flow does not in general remain steady; vice versa, we shall, in important cases, reduce non-steady flow to steady flow by changing to a suitable moving coordinate system.

A. Continuous Motion. Simple Waves.

59. Characteristics. The Mach angle. As stated, the characteristics C_+ and C_- in the x,y-plane depend on the specific flow under consideration. They are defined by the differential equation (see equations II(3),(5), Art. 9)

$$(10) \qquad (c^2 - u^2)dy^2 + 2uvdxdy + (c^2 - v^2)dx^2 = 0,$$

a relation obviously invariant under rotation of the coordinate system. Hence, without loss of generality, we may consider the flow at a point where the velocity component v vanishes, so that $|u| = q$, q being the speed of the flow. Then we obtain for the characteristic directions $\frac{dy}{dx}$ the relation $\frac{dy}{dx} = \frac{c}{\sqrt{q^2 - c^2}}$. We define the angle α by

$$(11) \qquad \sin\alpha = \frac{c}{q} = \frac{1}{M},$$

M being the Mach number of the flow. The angle α is often called the Mach angle. Then our result means that the characteristics C are the two sets of lines in the x,y-plane which intersect the streamlines at the Mach angle. (Occasionally characteristics in steady flows are called Mach lines.) Mach made the characteristics visible by roughening walls along which steady flow takes place, so that disturbances in the fluid (propagated along characteristics as seen in Art. 11,

Chapter II) are observed as lines which form the Mach angle with the wall, the latter being a streamline.*

The following is only another form of the definition of the Mach angle or of characteristics. In steady two-dimensional supersonic flow <u>the component of the flow velocity normal to the Mach lines is equal to the local sound speed c</u>.

Where the flow is subsonic, i.e., where $q < c_*$, the differential equations are elliptic. The solutions are then necessarily analytic and no characteristics can exist. Since, according to Art. , Chapter II, the Jacobian $u_x v_y - u_y v_x$ never vanishes for subsonic flow, the rôles of dependent and independent variables can be interchanged, and the differential equations (2) and (3') can be reduced to a system of linear differential equations. In this case our problem is essentially of the same type as problems in ordinary potential theory. In a steady subsonic flow there can be no subregion in which the state

---

* A geometrical construction of the Mach angle and the local sound velocity, due to Busemann [3] , is almost obvious. We draw an ellipse with $2c_*$ and $2\hat{q}$ as the minor and major axes respectively. To a given value q of the flow speed we find the point P on the ellipse at the distance q from the origin. Then the angle between OP and the major axis is the Mach angle α and the projection of OP on the direction of the minor axis is the local sound velocity c. The proof follows immediately from the form

$$(12) \quad \frac{q^2 - c^2}{\hat{q}^2} + \frac{c^2}{c_*^2} = 1$$

of Bernoulli's equation.

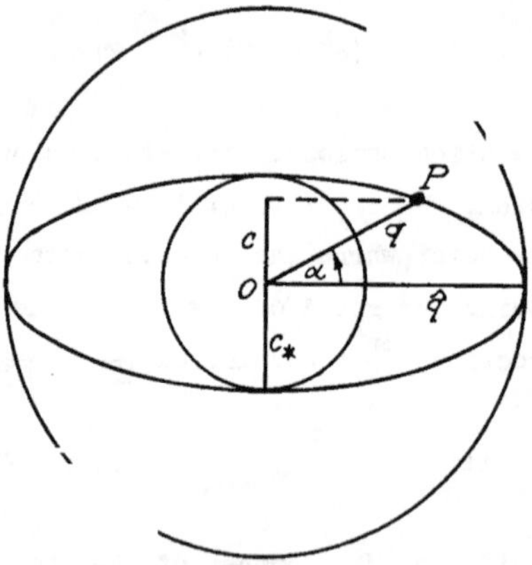

Figure 1
Busemann's geometric representation of flow speed q, sound speed c, and the Mach angle α.

is exactly constant, for the elliptic character of the differential equations in such a subregion and its neighborhood would imply analyticity of the solution, hence a constant state throughout. However, for supersonic flow a constant state in some parts is compatible with non-constant states in other parts, these different zones being separated by characteristics C.

We saw in Art. 16, Chapter II, that zones of non-constant states adjacent to zones of constant state are covered by simple waves,* i.e., are swept by straight characteristics (say C_+) along each of which u,v and consequently also the thermodynamic quantities p, ρ, T, c remain constant.

60. Simple waves. Epicycloidal shape of the characteristics Γ. The simple waves are linked to the characteristics Γ in the u,v-plane which satisfy the differential equation

$$(c^2 - u^2)du^2 - 2uvdudv + (c^2 - v^2)dv^2 = 0. \tag{13}$$

We recall from Art. 16, Chapter II, that the states in each simple wave are represented by the point of an arc of a characteristic and that, vice-versa, to an arc of any characteristic Γ there corresponds a simple wave. Hence by determining all possible simple waves we automatically solve the differential equation (13) and find the characteristics Γ. We shall see presently that the latter are the epicycloids generated by circles of diameter $\hat{q} - c_*$ rolling on the sonic circle $u^2 + v^2 = c_*^2$. These characteristics fill the annular ring between the sonic circle and the limit circle

---

* In one-dimensional flow the simple waves were rarefaction waves or compression waves, the latter leading to a shock. In two-dimensional steady flow continuous waves may equally well be expansion or compression waves. They were discovered by Prandtl and are frequently called Meyer-waves probably because "Meyer" is more readily pronounced than "Prandtl". (Meyer elaborated Prandtl's solution in his doctoral thesis [34]).

$u^2 + v^2 = \hat{q}^2$, leaving free the domain $u^2 + v^2 < c_*^2$, where the differential equations are elliptic, and $u^2 + v^2 > \hat{q}^2$, where the problem becomes meaningless. Through each point in the ring there pass two arcs of epicycloids, the branch which is orthogonal to characteristics C_- at corresponding points of the x,y-plane is superimposed being Γ_+, the other being Γ_-, according to (5), Art. 58.

As we saw in Art. 16, Chapter II, all possible simple waves are found by integrating the differential equations (1), (2), (3) of the flow under the additional condition that an arbitrarily prescribed family of straight lines in the x,y-plane consists of characteristics C, say a family of lines C_+. If these lines pass through a point E, the simple wave is called centered; otherwise we shall assume that the prescribed lines C_+ have an envelope E.

To determine the characteristics Γ, we consider the line C_+ through a point P and denote, as in Fig. 2, the length of the segment on C_+ from E to P by r, and by ω, the angle, measured counter-clockwise from the positive y-axis to the direction of C_+. The vector $\vec{q}$ of the flow velocity at P has a component L parallel to C_+ and N normal to C_+, and we denote by L and N the measures of these components, positive or negative according as the components of $\vec{q}$ point in the direction of increasing r or decreasing ω respectively, as in Fig. 2. We then have

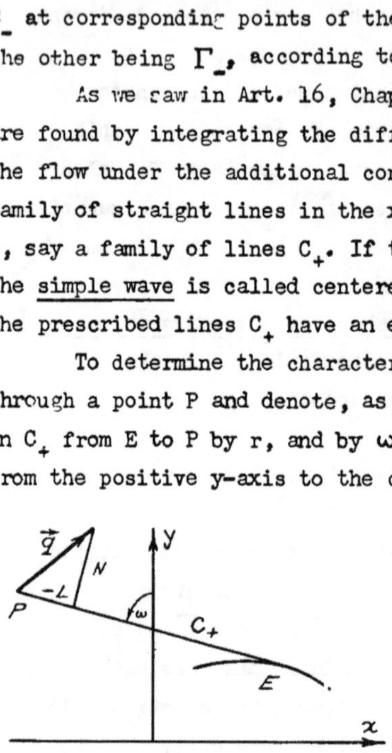

Figure 2
Indicating coordinates r, ω of a point and components L, N of flow velocity with reference to a characteristic C_+.

$$(14) \quad \begin{cases} u = -L \sin\omega + N \cos\omega \\ v = +L \cos\omega + N \sin\omega \end{cases}$$

and the continuity equation (3) and Bernoulli's equation (1) with reference to L and N as dependent and r and ω as independent variables can be shown to have the forms:

(15) $$(r\rho L)_r - (\rho N)_\omega = 0,$$

(16) $$\mu^2(L^2 + N^2) + (1 - \mu^2)c^2 = c_*^2$$

while from the equation $c^2 = A\rho^{\gamma - 1}$ we have

(17) $$\frac{2}{\gamma - 1}\frac{c_\omega}{c} = \frac{\rho_\omega}{\rho} .$$

Now we introduce the assumption that C_+ is a characteristic; hence, according to the preceding article, $N^2 = N^2(r,\omega) = c^2(r,\omega)$. Furthermore, we make use of the assumption that the solution to be determined is a simple wave, i.e., that all the quantities ρ,L,N are constant along each C_+ so that they depend not on r, but on ω alone. Then we have $\frac{\rho_\omega}{\rho} = \frac{2}{\gamma-1}\frac{N_\omega}{N}$, and by (15), $L = \frac{\rho_\omega}{\rho} N - N_\omega$; hence

(18) $$L - \frac{1}{\mu^2} N_\omega = 0.$$

Moreover, Bernoulli's equation becomes, by $c^2 = N^2$,

(19) $$N_\omega^2 + \mu^2 N^2 = \mu^2 c_*^2 .$$

The complete solution of the two equations for N and L is, with a constant ω_0 of integration

(20) $$\begin{cases} N = c_* \cos\mu(\omega - \omega_0), \\ L = -\frac{c_*}{\mu} \sin\mu(\omega - \omega_0), \end{cases}$$

whence we find, by (14), as the parametric representation of the characteristics,

$$(21)\quad \begin{cases} \dfrac{u}{c_*} = \dfrac{1}{\mu}\sin\mu(\omega-\omega_o)\sin\omega + \cos\mu(\omega-\omega_o)\cos\omega , \\ \dfrac{v}{c_*} = -\dfrac{1}{\mu}\sin\mu(\omega-\omega_0)\cos\omega + \cos\mu(\omega-\omega_0)\sin\omega . \end{cases}$$

From our construction it is obvious that all these curves result from one of them by rotation about the origin 0.

To identify the characteristics Γ as epicycloids between the circles $u^2 + v^2 = c_*^2$ and $u^2 + v^2 = \frac{1}{\mu^2}c_*^2 = \hat{q}^2$ we may, therefore, concentrate on the single curve

$$(22)\qquad \frac{N}{c_*} = \cos\mu\omega, \quad \frac{L}{c_*} = -\frac{1}{\mu}\sin\mu\omega, \quad \text{or}$$

$$(23)\qquad \begin{cases} \dfrac{u}{c_*} = \dfrac{1}{\mu}\sin\mu\omega\sin\omega + \cos\mu\omega\cos\omega , \\ \dfrac{v}{c} = -\dfrac{1}{\mu}\sin\mu\omega\cos\omega + \cos\mu\omega\sin\omega . \end{cases}$$

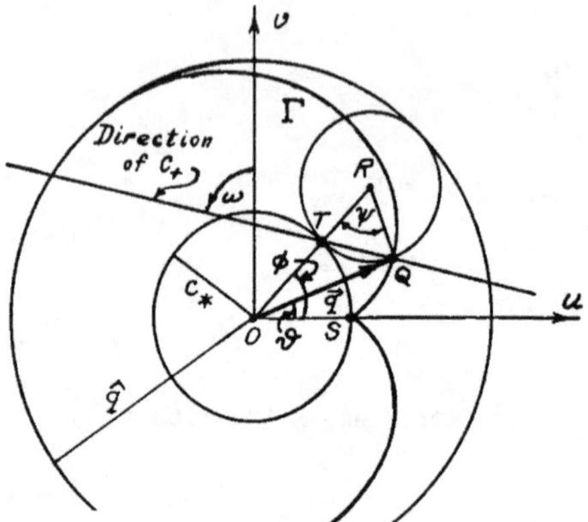

Figure 3
Construction of one pair of epicycloidal characteristics.

Let us, independently of the way in which this curve was obtained, consider an epicycloid described by the point Q on the circumference of a circle of radius $\frac{1}{2}\left(\frac{1}{\mu} - 1\right)c_*$ with center R rolling on the sonic circle $u + v = c_*^2$ with the center O, assuming that the initial position S of Q is at $u = c_*$, $v = 0$. With the notations indicated in Fig. 3 we have

$$(24)\quad \begin{cases} \frac{1}{2}\left(\frac{1}{\mu} - 1\right)\psi = \phi , \\ \frac{1}{2}\left(\frac{1}{\mu} + 1\right)\psi = \phi + \psi ; \end{cases}$$

and for the coordinates of Q.

$$
(25) \qquad \begin{cases} \dfrac{u}{c_*} = \dfrac{1}{2}\left(\dfrac{1}{\mu} + 1\right) \cos\phi - \dfrac{1}{2}\left(\dfrac{1}{\mu} - 1\right) \cos\,(\phi + \psi), \\[2ex] \dfrac{v}{c_*} = \dfrac{1}{2}\left(\dfrac{1}{\mu} + 1\right) \sin\phi - \dfrac{1}{2}\left(\dfrac{1}{\mu} - 1\right) \sin\,(\phi + \psi) . \end{cases}
$$

Introducing the auxiliary angle

$$
(26) \qquad \frac{\psi}{2\mu} = \phi + \frac{\psi}{2} = \omega \ ,
$$

we have $\phi = (1 - \mu)\omega$, $\phi + \psi = (1 + \mu)\omega$, and thus from (25)

$$
(27) \qquad \begin{cases} \dfrac{u}{c_*} = \dfrac{1}{2\mu}\left\{\cos\,(1 - \mu)\omega - \cos\,(1 + \mu)\omega\right\} + \dfrac{1}{2}\left\{\cos\,(1 - \mu)\omega + \cos\,(1 + \mu)\omega\right\}, \\[2ex] \dfrac{v}{c_*} = \dfrac{1}{2}\left\{\sin\,(1 - \mu)\omega - \sin\,(1 + \mu)\omega\right\} + \dfrac{1}{2}\left\{\sin\,(1 - \mu)\omega + \sin\,(1 + \mu)\omega\right\}, \end{cases}
$$

or

$$
(28) \qquad \begin{cases} \dfrac{u}{c_*} = \dfrac{1}{\mu} \sin\mu\omega\sin\omega + \cos\mu\omega\cos\omega \ , \\[2ex] \dfrac{v}{c} = -\dfrac{1}{\mu} \sin\mu\omega\cos\omega + \cos\mu\omega\sin\omega \ , \end{cases}
$$

which establishes our epicycloid as the characteristic Γ.

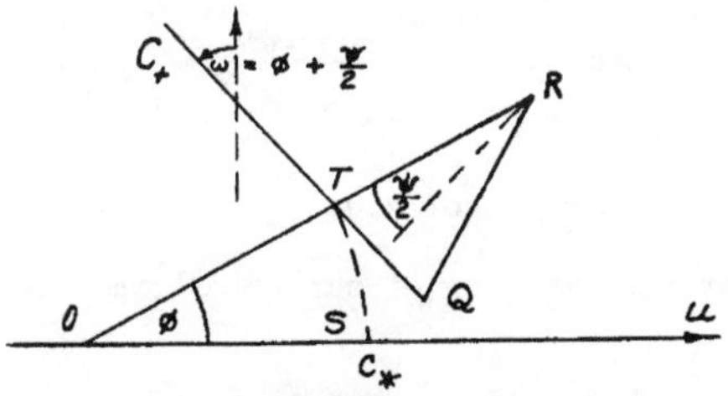

Figure 4
Construction of line C_+ and angle ω.

From Fig. 4 we infer immediately the geometric meaning of the angle ω as the angle between the line TQ and the v-axis. Since TQ is orthogonal to the epicycloid at Q, this confirms the previous definition of ω as the angle between the y-axis and the characteristic C associated with, and hence orthogonal to Γ. To $\psi = 0, \omega = 0$ corresponds the cusp S of the epicycloid; to $\omega > 0$, the upward arc, to $\omega < 0$ the downward arc; while $\psi = \pm\pi$ or $\omega = \pm\frac{\pi}{2\mu}$ give the points where the two arcs issuing from S touch the limit circle. These two arcs are called complete arcs; Γ_+ and Γ_- being, as remarked above, the arcs which are orthogonal to C_- and to C_+ respectively. All the other characteristics Γ are obtained from these two by rotation about O.

For the speed $q = \sqrt{L^2 + N^2}$ of the flow we obtain immediately the relation

$$(29)\quad \frac{q^2}{c_*^2} = \cos^2\mu\omega + \frac{1}{\mu^2}\sin^2\mu\omega = 1 + \left(\frac{1}{\mu^2} - 1\right)\sin^2\mu\omega = 1 + \frac{2}{\gamma - 1}\sin^2\mu\omega$$

$$= \frac{1}{\mu^2} - \left(\frac{1}{\mu^2} - 1\right)\cos^2\mu\omega\ ;$$

hence

$$(30)\qquad \sin^2\mu\omega = \frac{\gamma - 1}{2}\,\frac{q^2 - c_*^2}{c_*^2}$$

The angle ϑ between the vector $\vec{q}$ of the flow velocity and the positive x-axis is given (see (23)) by

$$(31)\qquad \tan\vartheta = \frac{v}{u} = \frac{\tan\omega - \frac{1}{\mu}\tan\mu\omega}{1 + \frac{1}{\mu}\tan\omega\tan\mu\omega},$$

Mach's angle α by

$$(32)\qquad \alpha = \text{arc sin}\left|\frac{c}{q}\right| = \text{arc sin}\,\frac{|N|}{|q|} = \text{arc sin}\,\frac{|\cos\mu\omega|}{1 + \frac{2}{\gamma - 1}\sin^2\mu\omega}\,,\ \text{or}$$

$$(33)\qquad \alpha = \text{arc tan}\left|\frac{N}{L}\right| = \text{arc tan}\,(|\mu\cot\mu\omega|).$$

We now enlarge upon the physical meaning of our geometric constructions.

A simple wave in the x,y-plane is called complete if it

corresponds to a complete arc Γ, i.e., if it is swept by a family of straight lines ranging over all directions normal to a complete arc Γ. If all these lines C pass through a point E the wave is called centered; but any family of lines obtained from the normals to an arc Γ by a deformation preserving the direction of each individual line is a possible set of lines C.

In a complete simple wave the flow is sonic at one end, while at its other end, the speed is the limit speed $\hat{q}$, pressure, density

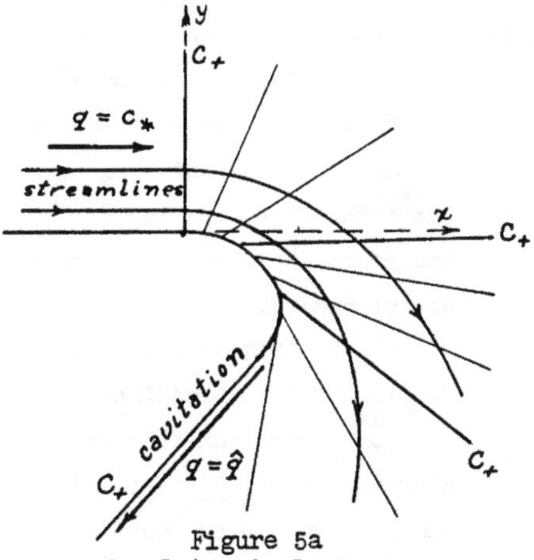

Figure 5a
Complete simple wave.

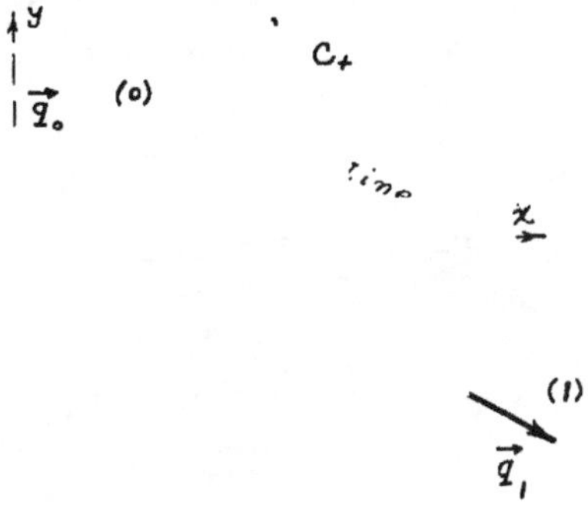

Figure 5b
Incomplete simple wave

and sound speed being reduced to zero; a zone adjacent to this end would therefore be a zone of cavitation.*

In general, a simple wave is <u>incomplete</u>, joining two regions of constant state, whose supersonic flow speeds q_0, q_1 and sound speeds c_0, c_1 satisfy the inqualitives $0 < c_1 < c_0 < c_* < q_0 < q_1 < \hat{q}$, the subscript ($_0$) denoting the region of higher density. Any simple

* Such a zone would not, of course, be an actual vacuum, but would be filled with a substance whose motion cannot be traced by the theory of fluid dynamics since the assumptions on which this treatment is based break down.

wave can be considered as a portion cut out of a complete wave, of which the two end parts then have no physical existence but are in reality replaced by constant supersonic states. We shall imagine any simple wave as being completed; then the direction of the characteristics C which have to be added are determined. In any simple wave, whether complete or not, we shall measure the angles ω, ϑ with reference to the sonic end of the completed wave, as described above.

Since the flow through a simple wave is reversible, the gas in a simple wave may flow either from the sonic end toward the cavitation zone or in the opposite direction. In the first case the wave is an expansion wave, pressure and density decreasing along the paths of the particles; in the second case we have a compression wave.

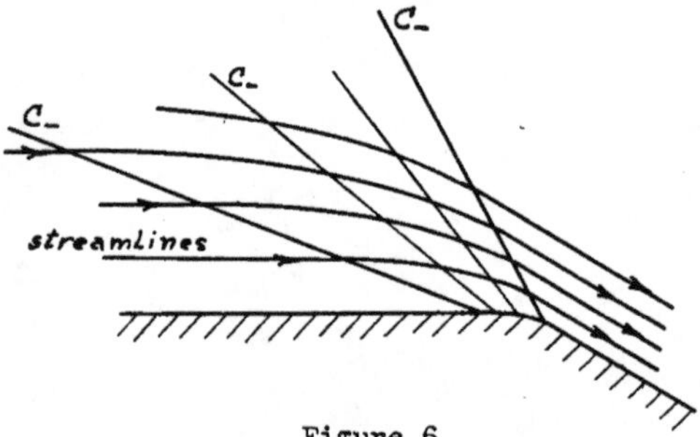

Figure 6
Simple compression wave.
Note that this flow is obtained by reversing the flow in Fig. 56.

It is easy to determine explicitly the streamlines and the conjugate, non-straight characteristics C_- (say) in the simple waves, using our parameters r and ω. Let us consider these curves in a complete simple wave (as in Fig. 5a). To find the streamlines take two points $P(r,\omega)$ and $Q(r + dr, \omega + d\omega)$ along a streamline. Then, from Fig. 7 it is evident that the streamlines are given by the differential equation

$$(34) \qquad \frac{dr}{-r d\omega} = \frac{L}{N} = -\frac{1}{\mu}\tan \mu\omega \ ,$$

see (20), or explicitly by

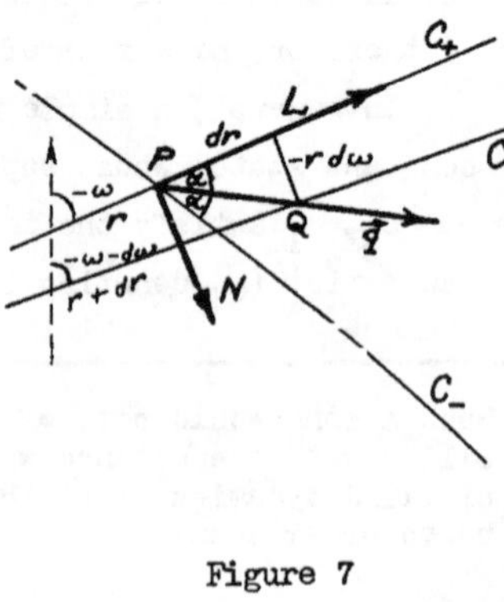

Figure 7

(35) $$r^{\mu^2} \cos\mu\omega = \text{const.}$$

By a similar construction one finds the differential equation of the curved conjugate characteristics C_- in a complete simple wave

$$\frac{dr}{-r d\omega} = \cot 2\alpha = \frac{\cos 2\alpha}{\sin 2\alpha} = \frac{1}{2}\left\{\frac{\cos\alpha}{\sin\alpha} - \frac{\sin\alpha}{\cos\alpha}\right\} = \frac{1}{2}\left\{\frac{L}{N} - \frac{N}{L}\right\} ,$$

since by Fig. 7, $\frac{L}{N} = \cot\alpha$; or, by (20),

(36) $$\frac{dr}{r d\omega} = \frac{1}{2}\left\{\frac{1}{\mu}\tan\mu\omega - \mu\cot\mu\omega\right\} ,$$

whose solution is the family of curves

(37) $$r^2 \sin\mu\omega(\cos\mu\omega)^{\frac{1}{\mu^2}} = \text{const.}$$

The characteristics C_- are all asymptotic to the two end characteristics C_+ of the complete wave, i.e., to $\omega = 0$ and $\omega = \frac{-\pi}{2\mu}$ for which we have $\vartheta = 0$ and $\vartheta = -\frac{1-\mu}{\mu}\frac{\pi}{2}$ respectively. The streamlines enter the simple wave with $\vartheta = 0$ and are also asymptotic to $\omega = -\frac{\pi}{2\mu}$, $\vartheta = -\frac{1-\mu}{\mu}\frac{\pi}{2}$.

Figure 8a shows a centered simple wave with streamlines, straight characteristics C_+ meeting in the center O, and curved characteristics C_-. This complete simple wave is bounded by two regions of constant state, one of sonic flow, the other of cavitation. Figure 8b shows an incomplete centered simple wave bounded by two constant states. Note that the incomplete wave corresponds to a sector (shown darker in Figure 8a) cut out of the complete wave, and that the states on either side of the incomplete wave are constant and that in the regions of constant state the straight characteristics C_+ are parallel and the conjugate characteristics C_- continue straight and parallel, both cutting the streamlines at a constant Mach angle.

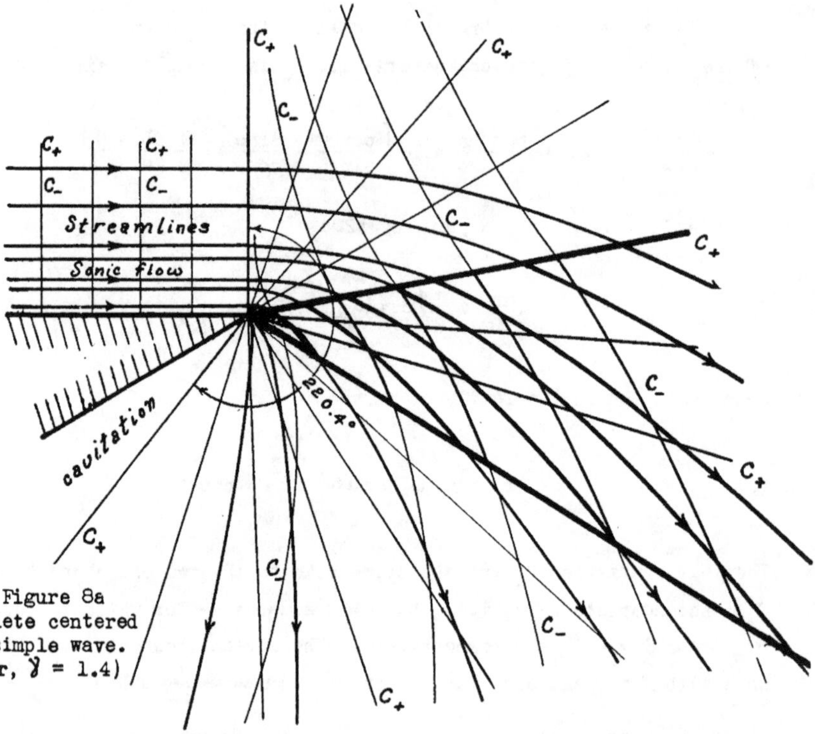

Figure 8a
Complete centered
simple wave.
(Air, $\gamma = 1.4$)

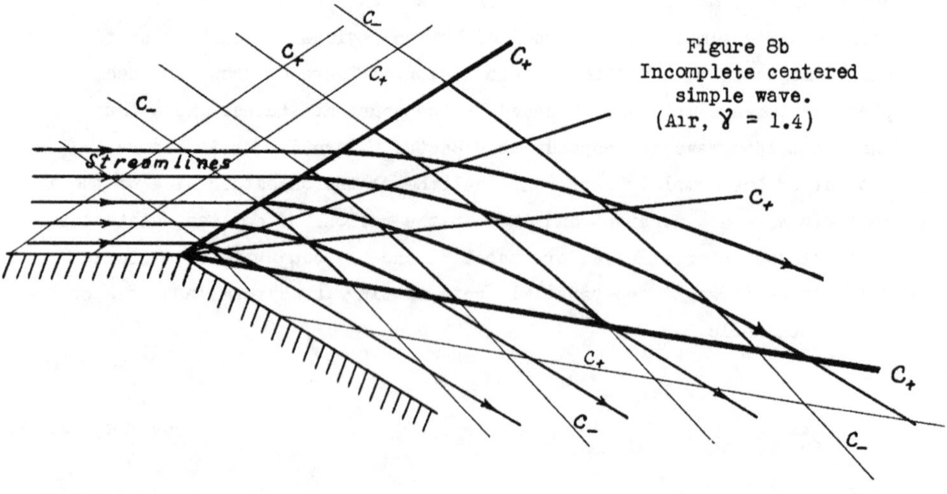

Figure 8b
Incomplete centered
simple wave.
(Air, $\gamma = 1.4$)

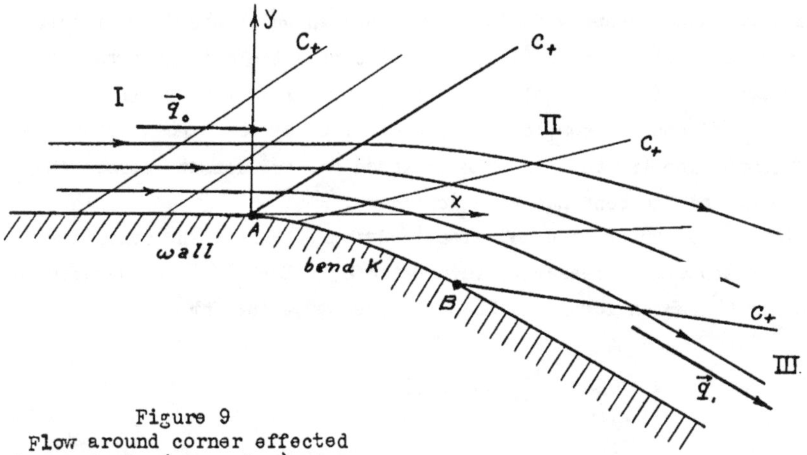

Figure 9
Flow around corner effected by a simple (expansion) wave.

61. Flow around a bend or corner. Construction of the simple waves.* The all-important instance of simple waves is in supersonic flow around a bend or sharp corner. We suppose the flow arrives with a constant velocity q_0 along a wall which is straight up to a point A, then bends along a smooth bend K from A to B and continues straight beyond the point B. We further assume that the oncoming flow is of constant state in a region adjacent to the straight part of the wall before A. Then the question is: How does the flow turn the corner? Or how will the flow continue along the bend K and along the straight wall beyond B?

If the oncoming flow is subsonic, the problem involves potential flow, governed by elliptic differential equations where the solution at any point depends on the boundary conditions even at remote points of the boundary.

We are, however, concerned with the case of supersonic flow. Then the solution is simpler. It can be obtained by piecing together various domains of the flow having essentially different analytic character,

* For flow around a projectile of polygonal shape see Epstein [35].

namely the zone (I) of constant state of the oncoming flow, a simple wave (II) which necessarily follows, and through which the flow effects its turn, and finally a zone (III) also of constant state which may be either a zone of flow parallel to the straight wall beyond B (if the simple wave effects the complete turn prescribed by the bend), or a zone of cavitation (if the flow has expanded to zero density before the full turn around the bend has been achieved).

Let us construct this solution in detail. First, the zone (I) of constant state is necessarily terminated by a Mach line or characteristic C_+^o or C_-^o which forms the Mach angle α_o, defined by

$$\sin \alpha_o = \frac{c_o}{q_o} , \tag{38}$$

with the direction of incoming flow, i.e., with the straight wall. This angle α_o is known since the state (o) or (I) and hence the corresponding sound speed c_o is known. Incidentally, the critical speed c_* of the flow is then given by

$$c_*^2 = \mu^2 q_o^2 + (1-\mu^2)c_o^2 .$$

Two positions of such a Mach line are possible, one inclined against the incoming flow and one inclined in the direction of this flow. For the moment we select the latter possibility, postponing the discussion of the former. To construct the adjacent simple wave (which is necessarily incomplete unless $q_o = c_*$) we have only to realize that at each point P of the bend the direction of the bend is that of the flow. We complete our simple wave backward beyond the initial characteristic C_+^o to the sonic end and refer the angle ω, which gives the direction of the straight characteristics C_+, and ϑ, which gives the direction of the flow, not as before to the y-, or x-axis but to the (imagined) sonic end of the corresponding complete simple wave, so that at this end $\omega = 0$, $\vartheta = 0$, and ω and ϑ decrease when q turns clockwise. Then at C_+^o we have values ω_o, ϑ_o determined either by the formulas of Art. 60, or simply by the following geometric procedure. We consider the complete epicycloidal arc Γ_-

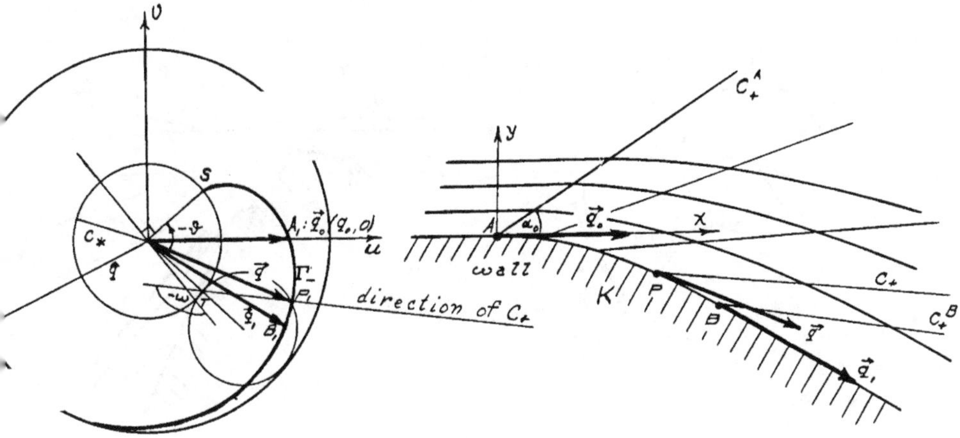

Figure 10
Construction of simple expansion wave in flow around a bend.

bending downward; on it the point A_1 at the distance q_0 from 0 corresponds to the beginning A of the bend. The angle at 0 subtended by the epicycloidal arc SA_1 is the initial angle ϑ_0, and likewise the point A_1 on Γ_- fixes ω_0. By rotating the diagram we bring the line OA_1 into a position parallel to the direction of the oncoming flow. Then to any point P on the bend K we obtain the corresponding position P_1 on Γ_- by drawing OP_1 parallel to the tangent on K at P. The characteristic C_+ through P is then determined as the line perpendicular to the direction Γ_- has at P_1, i.e., parallel to the line TP_1. Along C_+ the velocity $\vec{q}$ is parallel to the line OP_1 and the speed q is given by the length of OP_1.

If each point P of the bend has an image P_1 on Γ then the arc A_1B_1 of Γ_- represents the incomplete simple wave from which the flow emerges parallel to the straight wall beyond B with the speed equal to the length of the segment OB_1.

If, however, the bend is too strong, i.e., if the end B_1 of the arc Γ, where Γ touches the limit circle, corresponds to a point B of the bend

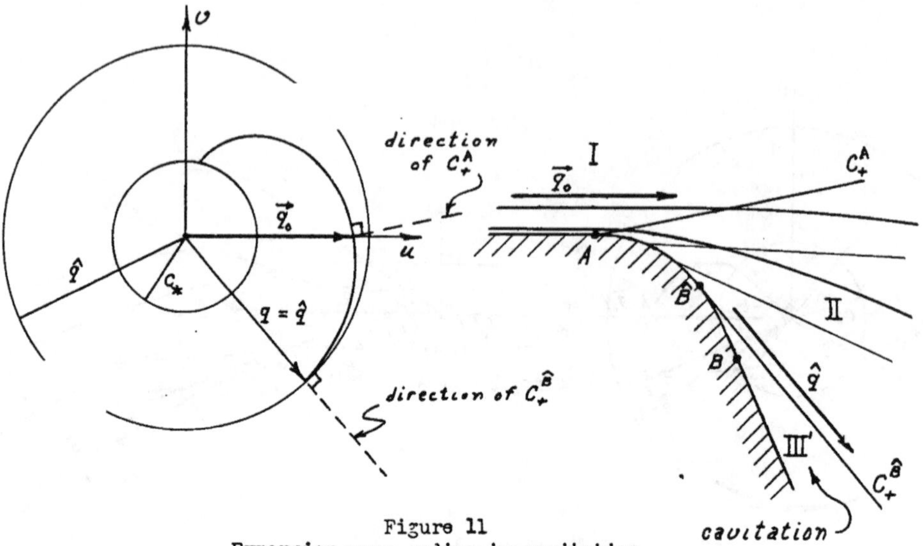

Figure 11
Expansion wave ending in cavitation occurring when bend is too strong.

between A and B, then the simple wave is completed by the characteristic $C_+^{\hat{B}}$ through $\hat{B}$, which is then tangent to the bend at $\hat{B}$; beyond this characteristic there will be cavitation (see footnote, p.163), and the flow in the wave zone (II) will acquire asymptotically the direction of this terminal characteristic (see preceding article).

By using the non-dimensional quantities $\frac{u}{c_*}$, $\frac{v}{c_*}$, $\frac{q}{c_*}$, μ, we can carry out this continuation with one single epicycloidal arc Γ^*, depending on μ only. It is obvious how to proceed graphically, for example, by using an arc Γ drawn on transparent paper.

One could, of course, proceed just as well analytically on the basis of the formulas of the preceding article. Instead, it is preferable for practical purposes to translate these formulas into a table for a complete simple wave, a table which depends only on γ or μ and which can be used in an obvious way, with proper interpolations if necessary. Such a tabulation is reproduced in Table I. To apply it in a specific case one determines the portion of the complete wave to be used from the ratio q_0/c_* or the Mach number $M_0 = q_0/c_0$ of the incoming flow. The table then

p/p_{max}	$\lvert\omega\rvert$	$\lvert\vartheta\rvert$	$M = q/c$	q/c_*
0.527	00°00'	0°00'	1.000	1.000
0.50	17°18'	0°25'	1.045	1.037
0.45	29°45'	1°57'	1.130	1.100
0.40	39°10'	4°06'	1.221	1.168
0.35	47°32'	6°44'	1.320	1.240
0.30	55°33'	9°52'	1.430	1.319
0.25	63°37'	13°34'	1.558	1.400
0.20	72°7'	17°58'	1.707	1.480
0.15	81°33'	23°23'	1.895	1.576
0.12	88°00'	27°21'	2.040	1.649
0.10	92°51'	30°28'	2.153	1.695
0.08	98°21'	34°07'	2.300	1.749
0.05	108°39'	41°15'	2.60	1.850
0.03	118°23'	48°17'	2.935	1.942
0.01	135°33'	61°14'	3.70	2.089
0.00	219°19'	129°19'	∞	2.440

Table I
Tabulation of quantities
characterizing a complete simple wave.*

(For air $\gamma = 1.405$)

gives the corresponding angles ω_0 and ϑ_0 and further quantities characterizing the appropriate simple wave can be read off next to the values of ϑ along the bend.

In the preceding article we saw that the ends of the complete arcs Γ, where Γ touches the limit circle, are given by $\omega = \pm\frac{\pi}{2\mu}$. Thus from (31), Art. 60, or from our diagrams, we see that the maximum angle through which the flow can be bent is

$$\hat{\vartheta} = \left(\frac{1}{\mu} - 1\right)\frac{\pi}{2} \tag{39}$$

* The tabulation p/p_{max} refers to the ratio of pressure to stagnation pressure $p_{max} = \frac{1}{.527}p_*$ (for air) see (9), Art. 58.

which is attained only in the ideal case of a complete wave, in which the flow starts at sonic speed c_* and ends at the limit speed $\hat{q}$. The following table gives numerical values for this maximum angle for various values of γ.

γ	1.00	1.20	1.25	1.30	1.40	1.67	2.00	3.00	7.00
$\hat{\vartheta}=\left(\frac{1}{\mu}-1\right)\frac{\pi}{2}$			180.0°	159.2°	130.4°	90.0°	65.9°	37.3°	13.9°

Table II
Maximum angle $\hat{\vartheta}$ through which a simple wave can turn a flow for various values of γ.

If $q_o > c_*$ then obviously the whole angle which the flow can possibly turn before cavitation occurs is correspondingly smaller.*

All the preceding discussion referred to a bend K with continuously turning tangent. The entire treatment, however, holds just as well for the idealized case where the gradual bend is replaced by a sharp corner K. Then the flow will arrive along the wall before K and suddenly turn at K into a new direction. The turn, discontinuous at the corner K, will be smoothed out into a continuous turn inside the wall; it will be effected by a centered simple wave, swept by a set of characteristics C all of which come from the center K. The theory of the preceding sections remains unchanged except that the angle indicating the direction of the flow along the bend K loses its meaning. Otherwise all the results subsist.

---

* For large values of $\frac{q_o}{c_o}$ or small Mach angles α one finds approximately as the maximum angle that the flow can turn the values

$$\left(\frac{1}{\mu^2}-1\right)\frac{c_o}{q_o} = \frac{2}{\gamma-1}\,\frac{c_o}{q_o}.$$

62. Compression waves. The simple waves considered in the preceding section are expansion waves; as is obvious from our formulas, density (and also pressure) decreases, while flow speed increases along the streamlines. However, compression waves in flow around a bend or corner are equally possible, as is immediately seen, for example, by considering the flow which is the reverse of an expansion wave. In the preceding sections we selected such solutions

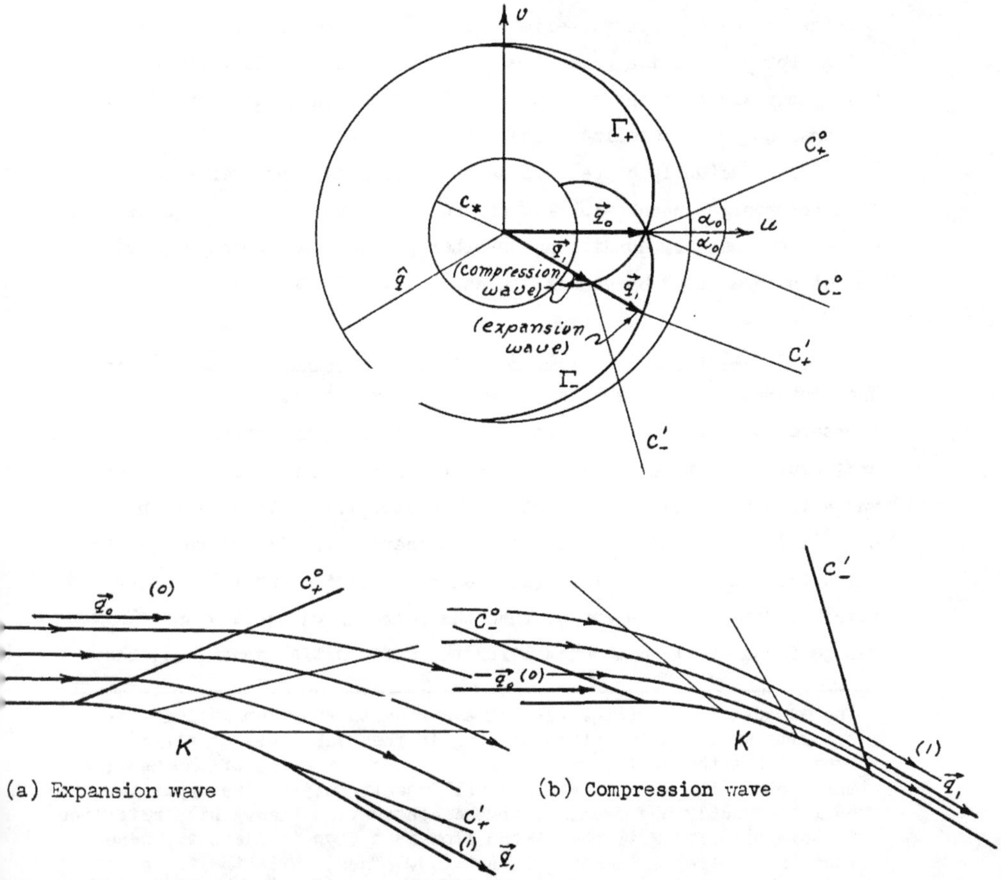

Figure 12
Construction of simple expansion wave (a) and compression wave (b) which can turn flow with oncoming velocity $\vec{q}$ around bend K.

of the differential equations as yield expansion waves along the bend K by choosing that branch of the epicycloid from the point A_1 in the u,v-plane which leads to larger values of $q^2 = u^2 + v^2$, and hence to smaller values of p and ρ. However, for a given bend or corner we might just as well have chosen the other epicycloidal branch through the point u_o, v_o in the hodograph plane, which corresponds to decreasing speed q and thus to increasing pressure and density. All our arguments and formulas remain essentially the same for the choice of a characteristic arc Γ representing a compression.

For compression waves the characteristics C_- along which u, v, ρ, c, p are constant are the Mach lines inclined, not towards, but against the streamlines, as indicated in the Fig. 12b.

What actually happens in an individual case, whether an expansion or a compression occurs in flow around a corner or a bend of a wall depends on boundary conditions on other parts of the boundary, and cannot be predicted by general rules.*

63. <u>Interaction of simple waves. Reflection on a rigid wall.</u> When interaction occurs between two simple waves, (I) and (II) (centered or not), we must exoect a situation as indicated in Fig. 14, analogous to that in the case of interaction of non-steady rarefaction waves in one dimension (see Art. 44, Chapter III). There will be a zone (III) of penetration bounded by a characteristic quadrangle, from which two simple waves (I'), (II') emerge. If the two interacting waves (I) and (II) are known, then the emerging waves (I') and (II') can be found easily without solving the differential equations; that

---

* The expansion wave (Fig. 12a) always results near the wall if the opposite wall remains plane as long as the Mach lines C_- starting from it hit the wall shown in Fig. 12 before the end of the bend; the contraction wave (Fig. 12b) will result only if the opposite wall is exactly a streamline of the contraction flow. With reference to this opposite wall the second flow is a flow in a concave bend (see the general discussion in Art. 65). Thus, while the two cases are complementary from a mathematical point of view, the first case occurs under more general conditions than the second.

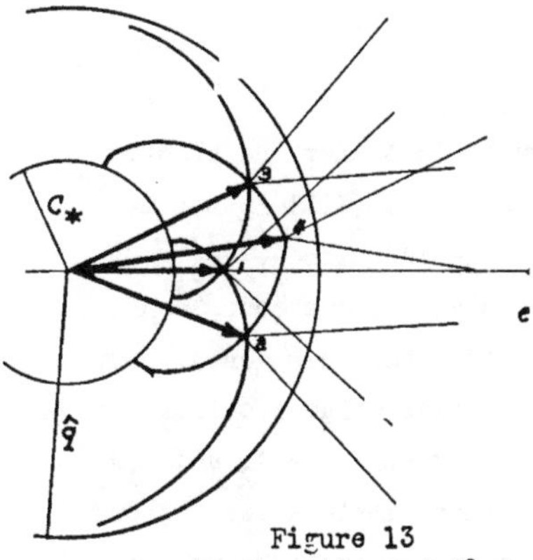

Figure 13
Construction of flow
in Fig. 14.

Figure 14
Interaction of two simple waves
showing region of penetration III.

is, we can determine the waves (I') (II') in the sense that the corresponding characteristic arcs Γ, or u and v as functions of the angle ϑ of the straight characteristics, can be found from our diagram (or from corresponding algebraic operations).

Suppose that in the domain (1) we have a constant supersonic velocity u_1, v_1, say $u_1 = q_1 > c_*$, $v_1 = 0$. Then in the hodograph plane the two waves (I), (II) are represented by two arcs of epicycloids 1-2 and 1-3 respectively. The waves (I'), (II') will again be represented by two epicycloidal arcs, as in the diagram, and the latter define as their intersection the point 4, representing the ultimate state of the fluid after the particles have passed both waves. The outcome of a reflection on a rigid wall simply corresponds to the interaction of two symmetric waves with the wall as the line of symmetry.

The preceding reasoning does not give detailed information about the width of the transmitted waves, the distribution of the straight characteristics C in them, or about the zone of penetration (III). To obtain such detailed information we have to determine the flow in zone

(III), and for this purpose we must solve an initial value problem for the differential equations of the flow. Zone (III) is not a simple wave, but represents a "general" state in which $\frac{\partial(u,v)}{\partial(x,y)} \neq 0$ (as can be seen by a detailed discussion, a one-to-one correspondence between the x,y-plane and the u,v-plane exists for this zone). Our task is the solution of a characteristic initial value problem. Zone (III) is reached along the two known characteristics, A_1A_2 and A_1A_3 (Fig. 14), which bound the end of the incoming simple waves. Along these initial characteristics the values of u and v are known corresponding to the two given arcs Γ which represent (I) and (II). For these characteristic initial data we must solve the hydrodynamical equations; thereby the two families of characteristics C covering (III) are determined, and in particular the characteristics A_2A_4 and A_3A_4. Beyond these curves simple waves (I') (II') will again occur, which can be immediately constructed, since at each point of these arcs u and v are determined, hence the slope of the straight characteristics C which sweep the simple waves (I') and (II') and carry constant values of u and v.

For the method of solving the initial value problem see the remarks in Arts. 10 and 13, Chapter II, and the literature quoted there.

Numerical or graphical integration is not difficult on the basis of this theory, and has been carried out in various cases.

64. Jets. Interaction of simple waves is the basis of a description of phenomena in a jet, formed by gas flowing with supersonic speed from an orifice into the atmosphere (see Fig. 15). For the moment we confine ourselves to a somewhat over-simplified theory proposed by Prandtl and Busemann (later in Art. 84, Chapter V we shall discuss a refined analysis of jets as they usually occur). On the basis of observation we suppose that the jet of escaping gas is separated from the quiet air at atmospheric pressure by a boundary wall consisting of a vortex layer (which becomes thicker along the jet and may ultimately consume

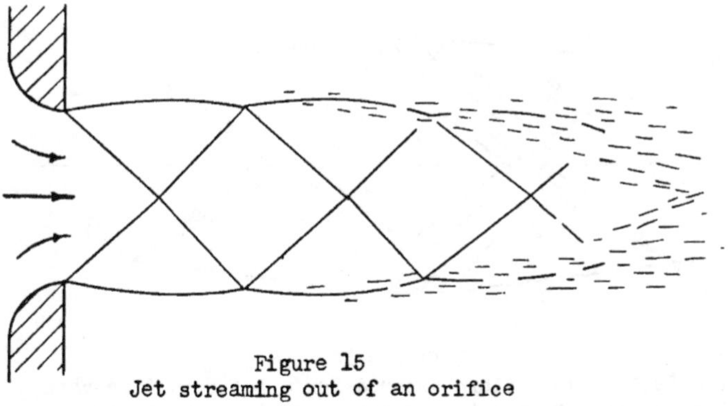

Figure 15
Jet streaming out of an orifice

it). Furthermore, we assume the phenomenon to be two-dimensional, steady, and isentropic, taking place, say, between two plane plates. Finally, we assume that the pressure p_0 in the oncoming parallel gas flow is greater than the atmospheric pressure p_A. Then Prandtl's description of the phenomenon as long as the jet is not yet destroyed by the boundary layer, is as follows.

At the corners of the orifice the compressed gas expands in two symmetrical centered simple waves to atmospheric pressure. These two simple waves interact and emerge again as simple waves from their zone of penetration. From the boundary layer which forms the wall of the jet the two simple waves are reflected again as simple waves which penetrate each other and continue as simple waves. Prandtl assumed that these waves converge in a center at the opposite side of the boundary.* These latter waves are contraction waves inasmuch as the gas flowing through them increases in density. As indicated in Fig. 16, the pattern is assumed to repeat itself, and to continue periodically if it were not for the influence of the boundary layer which gradually leads to a disintegration of the phenomenon. As seen in the preceding article, the pressures in the different regions of simple waves can be found directly from the known state of the

---

* It is rather certain that this assumption is not correct (see Art. 84, Chapter V).

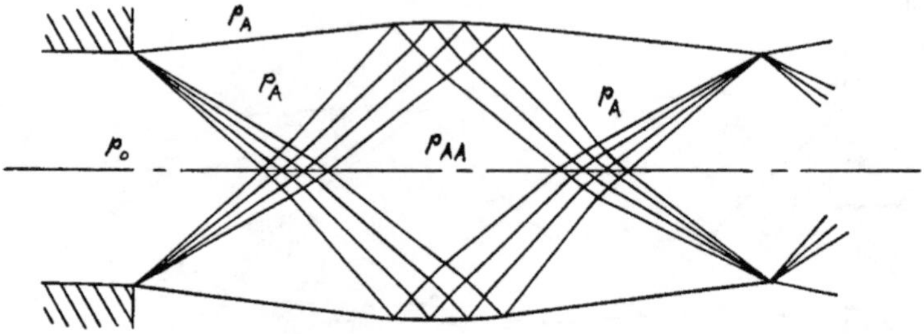

Figure 16
Prandtl's wave pattern assumed for a jet resulting when a parallel flow enters a region of lower pressure.

oncoming flow and the pressure p_A in the atmosphere by intersecting the epicycloids corresponding to the various simple waves. Likewise, it should be emphasized that the interaction of the first two expansion waves from the rim leads to a zone of constant pressure p_{AA} below the atmospheric pressure p_A.

In case the exhaust pressure p_o is less than the outside pressure p_A, a compression is needed for adjustment. According to Prandtl's and Meyer's theory this compression is effected by two symmetric <u>oblique shocks</u>, discontinuities which will be described in subsequent sections. These shocks will interact and lead to a zone of pressure $p_{oA} > p_A$. From there on the jet behaves as in the case $p_o > p_A$. (See the schematic representation in Fig. 17.)

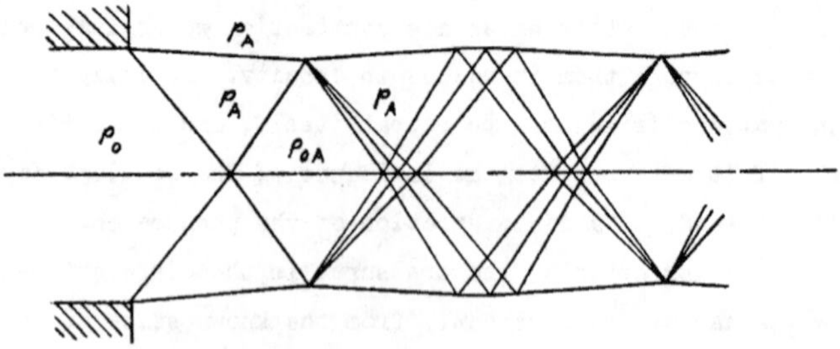

Figure 17
Wave pattern involving shock fronts assumed for a jet resulting when a parallel flow enters a region of higher pressure.

B. Oblique Shock Fronts

65. Flow in a corner. Oblique shocks. Qualitative description. Except for the remark at the end of the last section, we have assumed that velocity, pressure, density, etc., in our flows are continuous functions of x and y (though discontinuities in the derivatives may occur across characteristics or Mach lines). However, just as in the case of flow in one dimension it will often happen that a continuous flow is not compatible with the conditions of the problem. Then discontinuities are bound to occur. Fortunately, as in the case of one-dimensional flow, the simplest mathematical assumption, that of shock fronts, is in agreement with experimental evidence. The situation is similar to that in the one-dimensional case, where a simple wave was seen to result in a shock when it entails a contraction rather than a rarefaction and when the straight characteristics accordingly have an envelope inside the x,t-domain of the flow. Throughout our present analysis of two-dimensional steady flows the contrast between flow around a corner and in a corner plays a role corresponding to that of the contrast between expansive and compressive motion in the one-dimensional case. A continuous flow around a corner was seen to be possible in an expansion wave (and conceivably, under special conditions, in a contraction wave also). However, if a parallel supersonic flow arriving along a straight wall is forced to bend in a concave corner K a new situation arises.

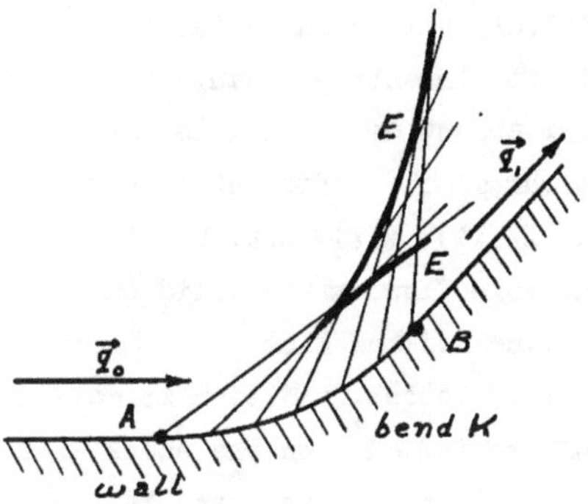

Figure 8
Envelope E of straight characteristics issuing from bend K.

In principle, our previous construction of a simple wave remains valid near the wall. There will be a characteristic (Mach line) through A in the x,y-plane, along which the constant flow (I) passes into a simple wave. But in contrast to the case of flow around a bend, the subsequent straight characteristics of the simple wave will now turn so that an envelope* originates inside the flow. The mathematically ambiguous state behind the envelope (where u, v would not be uniquely defined) is physically impossible. As observations indicate, it is avoided by a shock discontinuity, i.e., a line S of discontinuity for the quantities u, v, ρ, p, T, η. This shock line S will start with zero strength at the cusp of the envelope, and will run between the two branches of this envelope.

As we shall see presently, the shock conditions are such that particles crossing the shock front S from a zone of constant entropy will in general suffer different entropy changes, i.e., the flow ceases to be isentropic. Hence, generally speaking, behind a shock front consideration of non-constant entropy is unavoidable. However, in many important cases (the only ones that lend themselves to relatively simple analysis) the variation in entropy change is either absent or negligible so that our simple differential equations I (14), (17), Art. 3, of isentropic flow remain valid on both sides of the line S.

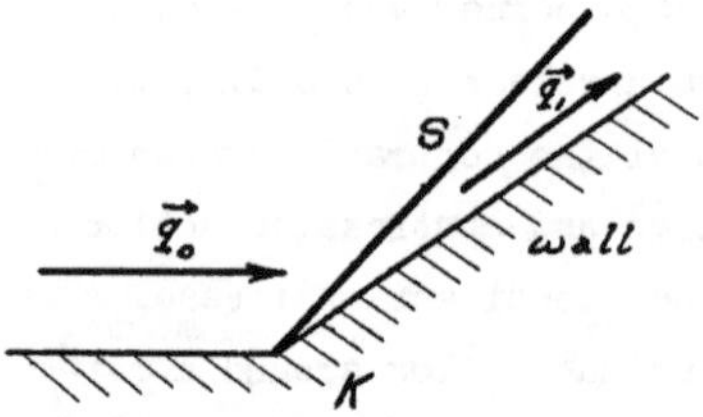

Figure 19
Straight shock line resulting from flow in sharp corner K.

This latter condition is certainly satisfied when the shock line S is straight and the state is constant on either side of S. Typical

* This envelope may degenerate into a single point, so that we have a centered simple wave.

for this situation and basically important in itself is the ideal limiting case where the bend K is concentrated in a sharp corner K if the flow arrives with constant supersonic speed at the corner K parallel to one leg of the angle and discontinuous by turns into the direction of the other leg, again at constant velocity. The sudden transition of direction and speed of the flow will then be effected across a straight shock line S extending from the corner K into the fluid obliquely to the direction of the flow. In this case the oblique shock front simply connects two zones (I) and (II) of constant state, and not only is there no complication from variations of entropy, but the algebraic character of the problem relieves us of concern with even the differential equations. The situation is qualitatively indicated in Fig. 19.

Incidentally, by reflection alone the straight part of the wall before the bend or corner, all the above remarks apply to two-dimensional supersonic flow against a wedge, which can be regarded as an idealized "projectile". Three-dimensional supersonic flow against the more realistic conical projectile will be treated later in Chapter V, Section B.

Before carrying out the quantitative analysis we must establish the general shock conditions.

66. Shock conditions in more than one dimension. Contact discontinuities.* For shocks in two dimensions (and in three dimensions as well) not restricted to steady isentropic flow, the discontinuity conditions are obtained from the principles of conservation of mass, of momentum, and of energy in exactly the same way as for one-dimensional flow.

An alternative method for obtaining the shock relations is to consider the flow from a moving coordinate system with respect to which the shock front is not oblique but normal so that the conditions of Chapter III can immediately be applied.

Before stating the shock relations a few remarks may be inserted.

* See Meyer [34].

First, by restricting our attention to a sufficiently small portion of the surface S of discontinuity we are justified in assuming S plane, and, in addition, by considering a sufficiently short interval of time, we may assume that the speed $\dot{\xi}$ with which the surface S moves in the direct of its normal is constant. Likewise, the two velocity vectors of the flow and altogether the two states (o) and (1) on the two sides of S may be assumed to be constant "in the small".

According to Galileo's principle of relativity, the shock conditions are to be invariant if referred to a coordinate system moving with a constant velocity relative to the original one. Thus, as stated, we may obtain these conditions by using the shock line S as one coordinate axis or, what is equivalent, by regarding the shock front as stationary, no matter whether or not the flow under consideration is steady. This leaves one more degree of freedom, inasmuch as we can move the origin of the coordinate system from which we measure our observed data with constant speed along the line S. Hence, without restriction of generality, we may assume the velocity component of the oncoming flow parallel to S as zero and thus visualize the flow as a flow of constant speed, meeting a stationary surface S of discontinuity at a right angle. If the speed q_o is not zero, i.e., if mass is transported through S, then the law of conservation of momentum requires that in the state (1) behind the shock the new velocity is likewise perpendicular to S. In other words, observed from a suitable coordinate system, an oblique shock front is always equivalent to a stationary one-dimensional shock front.

If, however, $q_o = 0$, seen from this coordinate system, then in the state (1) by the law of conservation of momentum, the normal component N of the velocity $\vec{q}$ vanishes while the tangential component may be arbitrary. In this case we have a contact discontinuity, described generally for an arbitrary surface S as follows. A contact discontinuity D is a surface D through which there is no mass flux, so that the flow is accordingly tangential seen from both sides, across which, however, density, pressure, temperature and entropy may be discontinuous. Such a contact surface may be considered as a vortex sheet, along which two different layers of the substance (or even of different substances) glide.

For genuine shock fronts through which there is a mass flux we distinguish (as in one dimension) between the front side and the back side of a shock front by saying that the fluid passes through the shock front from the front side to the back side.

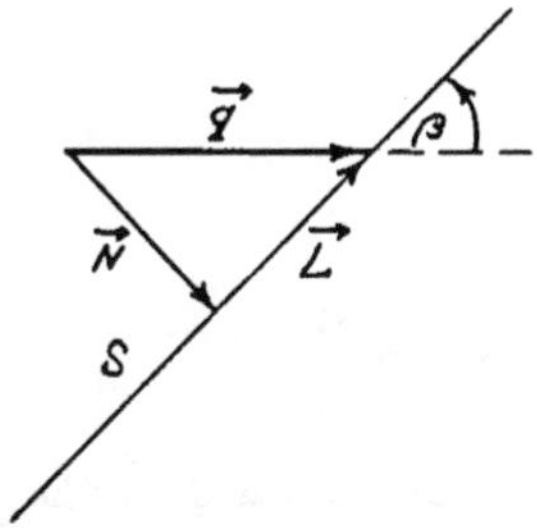

Figure 20

To formulate the shock conditions we describe by N the normal, and by L the tangential component of the flow vector $\vec{q}$, by $\vec{N}$ and $\vec{L}$ these components considered as vectors (see Fig. 20). Then the conditions are

Conservation of mass

(i') $$\rho_0(N_0 - \dot{\xi}) = \rho_1(N_1 - \dot{\xi}) = m$$

Conservation of momentum

(ii'_N) $$\rho_0 N_0(N_0 - \dot{\xi}) + p_0 = \rho_1 N_1(N_1 - \dot{\xi}) + p_1 .$$

Continuity of tangential component

(ii'_L) $$[L] = L_0 - L_1 = 0 .$$

In other words, the difference $\vec{q}_0 - \vec{q}_1$ of the velocity vectors is perpendicular to the shock line.

Conservation of energy

(iii') $$\tfrac{1}{2}\rho_0(N_0 - \dot{\xi})q_0^2 + \rho_0(N_0 - \dot{\xi})e_0 + N_0 p_0$$
$$= \tfrac{1}{2}\rho_1(N_1 - \dot{\xi})q_1^2 + \rho_1(N_1 - \dot{\xi})e_1 + N_1 p_1 \quad \text{with } q^2 = L^2 + N^2$$

Conditions for stationary shock fronts ($\dot{\xi} = 0$)

(i) $$N_0\rho_0 = N_1\rho_1 = m$$

(ii_N) $$\rho_0 N_0^2 + p_0 = \rho_1 N_1^2 + p_1 = P$$

(ii_L) $$L_0 = L_1$$

(iii) $$\frac{1}{2} q_0^2 + i_0 = \frac{1}{2} q_1^2 + i_1 = \frac{1}{2} \hat{q}^2$$

where $q^2 = N^2 + L^2$ is the square of the flow speed and $\hat{q}$ is the limit speed of the flow.

Of course, the latter form of the conditions remains valid, if N is replaced by $N - \dot{\xi}$, and thus a moving shock front introduced. The same relations (i), (ii_N), (iii) hold for contact discontinuities with $m = 0$; then $p_0 = p_1$, $N_0 = N_1$ follows, while now L_1 and L_0 may be different as well as ρ_0 and ρ_1. In what follows, we shall consider genuine shocks with $m \neq 0$, $L_0 = L_1$, unless the contrary is explicitly stated.

A slightly different and more symmetric way of writing the first two conditions for stationary shocks, expressing the conservation of the mechanical quantities mass and momentum, is the following:

(ii_0^*) $$\frac{p_1 - p_0}{\rho_0} = \vec{q}_0(\vec{q}_0 - \vec{q}_1) ,$$

(ii_1^*) $$\frac{p_0 - p_1}{\rho_1} = \vec{q}_1(\vec{q}_1 - \vec{q}_0) \text{ ; from which}$$

(ii^*) $$(p_1 - p_0)(\tau_0 - \tau_1) = (\vec{q}_0 - \vec{q}_1)^2$$

follows. (ii_L), (ii_o^*) and (ii_1^*) together are equivalent to the equations (i) and (ii_N).

As in the case of one-dimensional shocks it is important to observe that the conditions (i), (ii_N), (ii_L), (ii_o^*), (ii_1^*) are valid on the basis of mechanics alone. The thermodynamical characteristics of the medium intervene only through condition (iii). (As previously stated (Art. 33, Chapter III), there are many instances where the solution of flow problems is greatly eased by the possibility of disregarding the thermodynamical shock condition.)

67. Conditions for shocks in polytropic gases. As for one-dimensional flow the thermodynamical shock condition (iii) in the case of a polytropic gas (with which we shall be primarily concerned) is, since $i = \frac{\gamma}{\gamma - 1}\frac{p}{\rho} = \frac{1}{\gamma - 1} c^2$,

$$(iii_\gamma) \qquad \mu^2 q_0^2 + (1 - \mu^2)c_0^2 = \mu^2 q_1^2 + (1 - \mu^2)c_1^2 = c_*^2$$

where c^2 is the sound speed, $c_* = \mu\hat{q}$ the critical speed, and $\mu^2 = \frac{\gamma - 1}{\gamma + 1}$. Likewise, as in the case of normal shocks, we may replace (iii) by the relation (see III (49), Art. 36):

$$(40) \qquad \frac{p_1}{p_0} = \frac{\mu^2 - \frac{\rho_1}{\rho_0}}{\mu^2 \frac{\rho_1}{\rho_0} - 1},$$

which connects pressures and densities only and is invariant under translations.

Of great importance for the study of shocks in gases is the generalization of Prandtl's relation $q_0 q_1 = c_*^2$ previously obtained for normal shocks (Art. 35, Chapter III). To apply this relation in the case of an oblique shock, we write

$$(41) \qquad \vec{q} = \vec{N} + \vec{L}$$

where $\vec{L}$ is the vectorial component of the flow parallel, and $\vec{N}$ the vectorial component normal to the stationary shock line S, so that $\vec{N}\,\vec{L} = 0$. By substituting in er i' equation $\mu^2 q^2 + (1 - \mu^2)c^2 = c_*^2$ we find $\mu^2 N^2 + \mu^2 L^2 + (1 - \mu^2)c^2 = c_*^2$, or

$$(42) \qquad \mu^2 N^2 + (1 - \mu^2)c^2 = c_*^2 - \mu^2 L^2 = \tilde{c}_*^2$$

where $\tilde{c}_*$ is the critical speed in a new coordinate system in which the shock front is normal. Hence by the previous result (III (iii_p), Art. 35),

$$(43) \qquad N_o N_1 = \tilde{c}_*^2$$

or

$$(iii_p) \qquad N_o N_1 = c_*^2 - \mu^2 L^2$$

an important relation, which may be used instead of the preceding forms of the thermodynamical shock condition.

Some general and significant information follows from the relations developed so far. Equation (iii_p) shows that

$$(44) \quad q_o q_1 = \sqrt{N_o^2 + L^2}\,\sqrt{N_1^2 + L^2} > N_o N_1 + L^2 = c_*^2 + (1 - \mu^2)L^2 > c_*^2$$

because of $\mu^2 < 1$. Hence, if (o) is the front side of the shock and therefore $N_o > N_1$, then $N_o > c_*$; but we cannot, as for normal shocks, conclude that $N_1 < c_*$.

(a) <u>The speed of a flow on the front side of a shock (observed from the shock front) is supersonic; the speed on the back side may be subsonic or supersonic.</u> (The latter possibility becomes obvious if we recall that the tangential component L is quite arbitrary and may be chosen larger than c_*, whereupon q_o and q_1 both become larger than c_*.

(b) An even more immediate remark is: By an oblique shock in a gas the flow direction will always be turned towards the shock line S. For, the normal component N will become smaller when the flow crosses S, while the tangential component L remains unchanged.

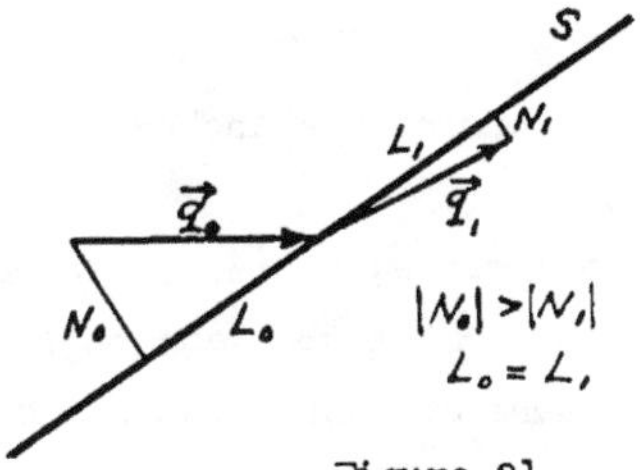

Figure 21
Showing how flow is
turned toward the shock line.

(c) A further remark concerns the relative position of Mach lines (characteristics) and flow vectors. For a normal shock line S facing towards (o) we have in the supersonic zone (o) two directions of Mach lines C_+ and C_-, the Mach angle α_0 being defined by $\sin\alpha_0 = \frac{c_0}{q_0}$. In the zone (1), however, the flow is subsonic; consequently there are no Mach lines in the zone behind the normal shock front (see Fig. 22).

For oblique shocks the latter statement remains true as long as $q_1 < c_*$, i.e., as long as the state (1) is subsonic. If $q_1 = c_*$ a Mach line will appear perpendicular to the flow velocity in (1). If $q_1 > c_*$ we have two different Mach line directions in (1). It is important to realize that their position is as indicated in Fig. 23.

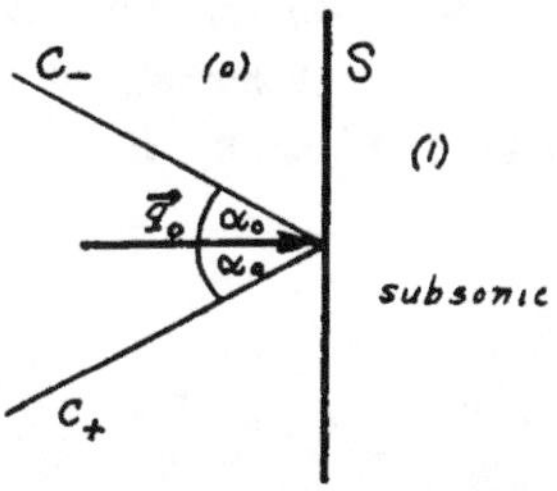

Figure 22
Characteristics C exist
only in front of
a normal shock line.

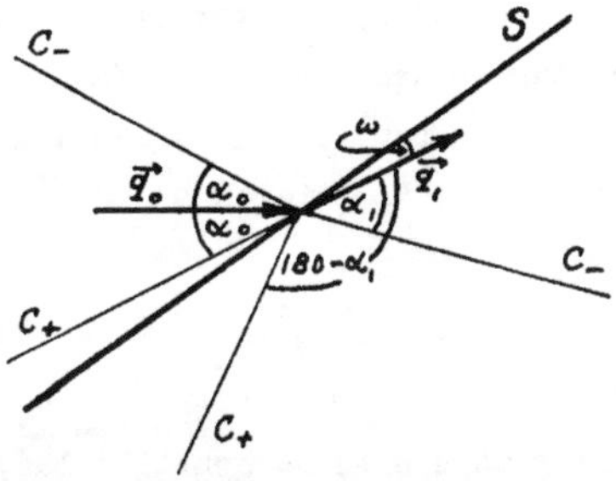

Figure 23
Positions of characteristics C
in front and in back of
an oblique shock line.

(Note that the sum of the angles which C_+ and C_- make with q_1 equals 180°)

In other words, behind the shock front the two Mach directions (if they exist) forming the same angle α_1 with the flow direction will lie in the obtuse angle between S and the direction of the flow vector $\vec{q}_1$. The statement is obviously equivalent to $\omega < \alpha_1$ where ω is the acute angle between the flow in (1) and the line S. Now we have $\sin\alpha_1 = \frac{c_1}{q_1}$, $\sin\omega = \frac{N_1}{q_1}$; hence we have to prove $c_1 > N_1$. This, however, follows immediately when we reduce the oblique shock to a normal shock by referring the motion to a coordinate system moving with the velocity $\vec{L}$. Then the normal components N_0, N_1 of the flow velocities remain unchanged, as do the sound speeds c_0, c_1, which depend only on pressures and densities. In the new coordinate system the speed of the flow behind the shock front is N_1, and since the shock front is normal, the speed behind it is subsonic, i.e., $N_1 < c_1$.

(d) Finally we observe that a <u>weak shock</u> for which the difference $q_1 - q_0$, or what amounts to the same thing, the difference $\rho_1 - \rho_0$ or $p_1 - p_0$, is small, differ but little from a <u>sonic disturbance</u>. That is, for a sequence of shocks for which $\frac{p_1}{p_0}$ tend to unity, the direction of the shock line S will tend to a Mach direction or characteristic direction (compare the similar station for shocks in one dimension described in Art. 35, Chapter III). As we have seen, the characteristic direction can be defined by the fact that the component N normal to it is equal to the sound speed c. Now, if in the limit $N_0 = N_1 = N$, the basic relation (iii$_p$) states $N^2 = c_*^2 - \mu^2 L^2$ or

$$c_*^2 = (1 - \mu^2)N^2 + \mu^2(L^2 + N^2) = (1 - \mu^2)N^2 + \mu^2 q_0^2 \quad .$$

On comparing this with Bernoulli's law, $(1 - \mu^2)c^2 + \mu^2 q_0^2 = c_*^2$, we find indeed $c^2 = N^2$.

68. Geometrical representation of transitions through oblique shocks in two dimensions. In Art. 39, Chapter III, we have interpreted graphically the transition through a stationary normal shock. In two dimensions we obtain a similar representation by a surface $p(q)$ for $q^2 = u^2 + v^2$ in a three-dimensional u, v, p-space. We simply rotate the curve $p(u)$ * of Art. 39 about the p-axis and obtain the pressure surface. $p = p(q) = p(u,v)$, where $q^2 = u^2 + v^2$. $p = p(q) = p(q,\eta)$ depends on the entropy η, decreasing to zero when for fixed q the entropy increases to infinity. As before we have

$$(46) \qquad p_u = -\rho u , \qquad p_v = -\rho v$$

$$(47) \qquad p_{uu} = -\rho\left(1 - \frac{u^2}{c^2}\right) , \quad p_{vv} = -\rho\left(1 - \frac{v^2}{c^2}\right), \quad p_{uv} = \frac{uv}{c^2}\rho ,$$

$$(48) \qquad p_{uu}p_{vv} - p_{uv}^2 = \rho^2\left(1 - \frac{q^2}{c^2}\right) \therefore$$

Thus the curvature of the pressure surface is positive for subsonic speed, $q^2 < c^2$, and negative for supersonic speed, $q^2 > c^2$. The two parts of the surface may be called the "dome" and the"rim", the latter ending horizontally at $q = \hat{q}$ (see [3]).

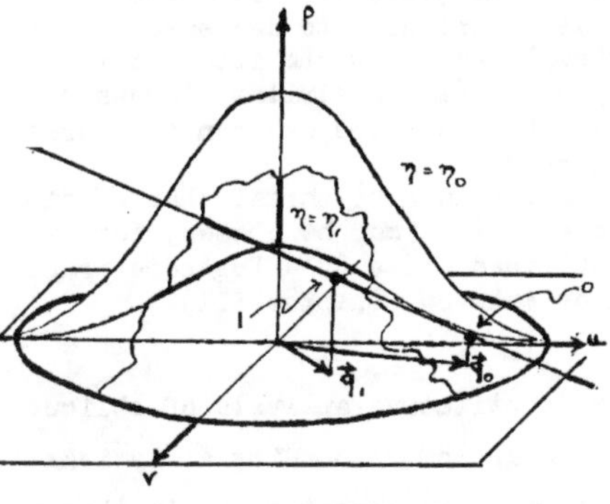

Figure 24
Representation of shock transition by u,v,p-surfaces.

While a single pressure surface represents various states of the gas which may belong to different points in the same continuous steady flow, two states u_0, ρ_0, p_0 and u_1, ρ_1, p_1 on the two sides of an oblique shock will be represented by two points of two different pressure surfaces with the same $\hat{q}$ but with different entropies η_0 and η_1. As is easily (see Art. 39) seen from our shock conditions, the straight line in the u, v , p-space

* Explicitly, for a polytropic gas with $p = A(\eta)\rho^\gamma$, we have

$$(45) \qquad p = A^{\frac{1}{1-\gamma}} \left[\frac{\gamma - 1}{2} (\hat{q}^2 - u^2 - v^2) \right]^{\frac{\gamma}{\gamma - 1}} .$$

connecting two points u_o, ρ_o, p_o and u_1, ρ_1, p_1 is tangent to the two surfaces with the entropies η_o and η_1.

We now consider a fixed state (o) and ask for all states (1) that can be connected with (o) by an oblique shock. The straight lines in the u,v,p-space connecting (o) with all such states (1) will then lie on the plane T_o tangent to $p = p(u,v)$ at the point (o). Let us consider a surface $p = p(u,v)$ with any entropy and intersect it with the tangent plane T_o. Through the point (o) there will in general be two lines tangent to the intersection curve. The points of contact of these tangents with the intersection curve represent states (1) that can be connected with (o) by a shock.

To investigate the locus of the possible points (1) we assume that the state (o) is supersonic, $q_o > c_o$, i.e., that the point (p_o, u_o, v_o) is on the rim. We then consider the family of surfaces $p = p(u,v)$ for decreasing entropy. When the entropy is infinite p is identically zero. When the entropy increases the dome of the surface $p = p(u,v)$ will rise, touch, and eventually cross* the tangent plane T_o. The state at which the dome touches T_o corresponds to a normal shock. After the dome has crossed the plane T_o the intersection will be an oval and there will be two tangent lines yielding two states (1) that correspond to two possible oblique shocks. When the entropy decreases to η_o these two points move into the point (o). So far the state (o) was on the front side, the state (1) at the back side of the shock. After η becomes less than η_o this situation is reversed. The state (1) will then have the greater speed, $q_1^2 > q_o^2$ and, as η continues to decrease, the point representing (1) will eventually approach the intersection of T_o with the limit circle, i.e., $q_1^2 \to \hat{q}^2$. (A similar discussion could be carried out for $q_o < c_o$; the state (o) is then of course always on the back side of the shock.)

It may be added that , as in the case of normal shocks, other geometric three-dimensional representations may be chosen, for example, a generalization of that obtained above by a Legendre transformation (see discussion at end of Art. 39, Chapter III).

69. The shock polar. For a quantitative analysis of oblique shocks constructions in two dimensions and corresponding analytical methods are by far preferable to graphical representations in three-dimensional space.

---

* At the intersection of the rim with the plane T_o there are no tangent lines through (o) as long as the entropy is greater than η_o.

Instead of obtaining such a two-dimensional construction by projection of the pressure surface of Art. 68 on a plane we proceed directly from our analytical shock conditions.

We shall discuss oblique shocks for flow in an x,y-plane, u and v being the components of the flow velocity. The shock line S is assumed to be straight and the states (o) and (1) on both its sides to be constant, these assumptions being always valid "in the small".* We assume the shock to be stationary and the medium to be polytropic gas. The shock direction may be characterized by the angle β between the shock line S and the direction of the oncoming flow. The angle between the vector $\vec{q}_o$ of the oncoming and $\vec{q}_1$ of the outgoing flow will be denoted by θ. Placing the coordinate system so that the oncoming flow is parallel to the x-axis and using our previous notation, N and L, for the measures of components of $\vec{q}$ normal and tangential to the shock line S respectively, we have

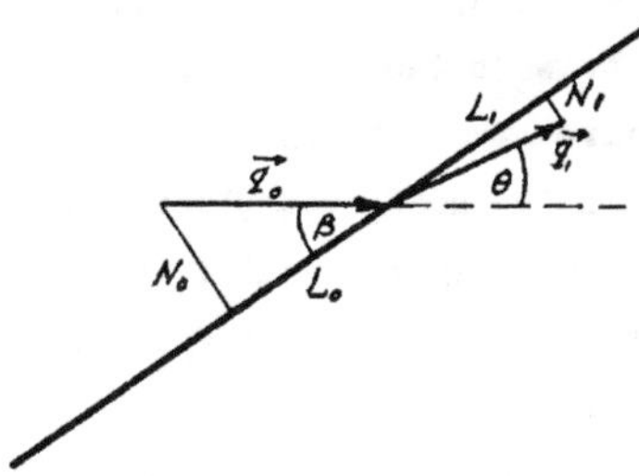

Figure 25

$$
(49) \quad \begin{cases} u_o = q_o \quad , \quad v_o = 0 \\ u_1 = L_1\cos\beta + N_1\sin\beta \quad , \quad v_1 = L_1\sin\beta - N_1\cos\beta \\ L_o = L_1 = q_o\cos\beta, \quad N_o = q_o\sin\beta \end{cases}
$$

* We recall that it is sufficient to study motion in an x,y-plane; for, if seen in a small neighborhood of a point P on S the flow can be considered as a two-dimensional flow in the plane orthogonal to S and containing the velocity vector of the oncoming streamline, and also the outgoing streamline (since the two velocity vectors have a common tangential component. Locally, therefore, all oblique shocks are phenomena in a plane, except at singular points of S such as the vertex of a cone.

Now Prandtl's basic relation (iii_P), Art. 67, yields $N_1 = \dfrac{c_*^2 - \mu^2 L^2}{N_0}$ and thus, we obtain the following relations:

$$(50) \qquad \begin{cases} u_1 = (1-\mu^2) q_0 \cos^2\beta + \dfrac{c_*^2}{q_0} \\ v_1 = -\left\{ \left[(1-\mu^2)\cos^2\beta - 1 \right] q_0 + \dfrac{c_*^2}{q_0} \right\} \cot\beta \,. \end{cases}$$

These equations show that for a given critical speed c_* and given speed q_0 of the oncoming flow the angle β between the shock line S and the direction of the vector $\vec{q_0}$ determines the outgoing flow velocity $\vec{q_1}$. If β varies the point u, v in the hodograph-plane with the rectangular coordinates u and v describes a curve, called the shock polar which corresponds to the value q_0 and to the critical speed c_*, and which is represented by (50) with β as a parameter.

The shock polar is the well-known "Folium of Descartes",* with a double or isolated point at $u = q_0$, $v = 0$ and with an asymptote $u = (1-\mu^2) q_0 + \frac{c_*^2}{q_0} = U$ (this asymptote lying outside of the limit circle $u^2 + v^2 = \hat{q}^2$.

Of the shock polar only the part with $u^2 + v^2 = \hat{q}^2$ has physical significance. It is the geometrical locus of all points (u_1, v_1) characterizing a state (1) which can be connected with the fixed state (o) by a stationary shock.**

Introducing

$$\tilde{u} = \frac{c_*^2}{q_0}$$

(the speed of the flow that would result from a normal shock), eliminating β, and writing u, v, q instead of u_1, v_1, q_1 we obtain for the shock

---

* Busemann [3] calls this curve a "strophoid" in disagreement with custom; for the sake of brevity we shall follow Busemann in using the word strophoid.

** c_* is known when the state (o) is given, from $\mu^2 q_0^2 + (1-\mu^2) c_0^2 = c_*^2$.

polar the equation

$$v^2 = (u - q_0)^2 \frac{u - \tilde{u}}{U - u} \quad . \tag{51}$$

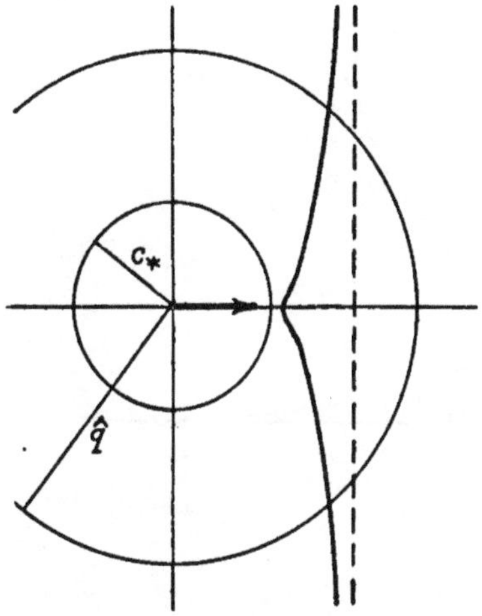

Figure 26a
Shock polar for
$q_0 < c_*$.

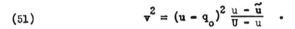

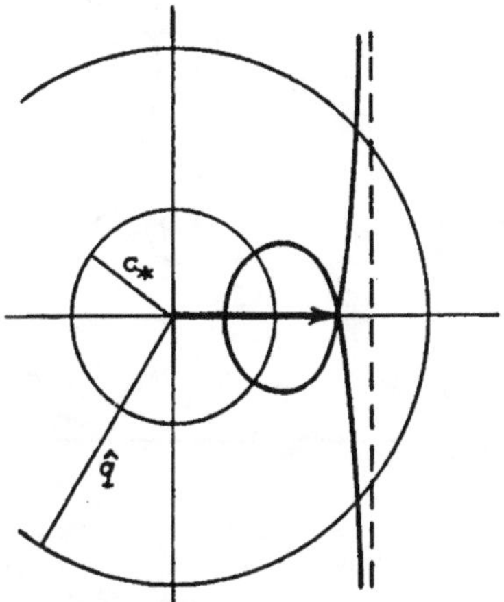

Figure 26b
Shock polar for
$q_0 > c_*$.

The one-parametric family of possible shock transitions from the state (o) can also be illustrated by a shock polar in the θ,p-plane, exhibiting the pressure on the other side of the shock and the angle θ through which the flow is turned (again the state (o) is assumed to be fixed, e.g., by prescribing q_0, ρ_0, p_0 or q_0, p_0, c_*). In parametric form with β as parameter we find from (ii_0^*), Art. 66 and (50) the equations

$$(52) \quad \begin{cases} p - p_0 = \rho_0 \left\{ q_0^2 [1 - (1 - \mu^2)\cos^2\beta] - c_*^2 \right\} \\ \tan\theta = \dfrac{v}{u} = \left\{ 1 - \dfrac{1}{(1 - \mu^2)\cos^2\beta + \left(\dfrac{c_*}{q_0}\right)^2} \right\} \cot\beta \end{cases}$$

which define the image of our strophoid in the θ, p-plane; this θ,p-shock polar is shown in Fig. 27

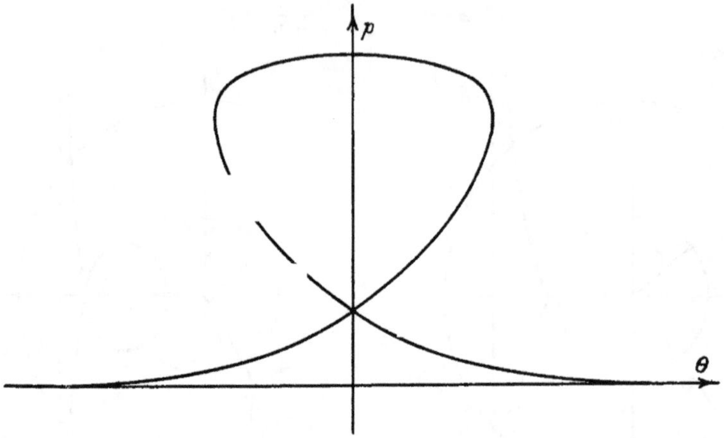

Figure 27
Shock polar in the θ,p-plane.

70. Discussion of oblique shocks by means of shock polar. The geometry of the shock polar gives immediately information about the quantitative situation in shock transitions. The shock polar, which depends on c_* and q_0 as parameters, can be used to construct oblique shocks as follows. Fig. 28 shows a strophoid for $q_0 > c_*$ with the double point at P_0, the endpoint of the vector $\vec{q}_0$.

To any point P on the shock polar we find the velocity behind the shock front as the vector OP. The angle θ through which the shock turns the incoming flow is the angle between OP and the u-axis. The direction of the shock line S, making the angle β with the incoming flow, is perpendicular to the line connecting the double point P_0 with the point P, in accordance with the fact stated before that the vector $\vec{q} - \vec{q}_0$ is perpendicular to the shock line. The components

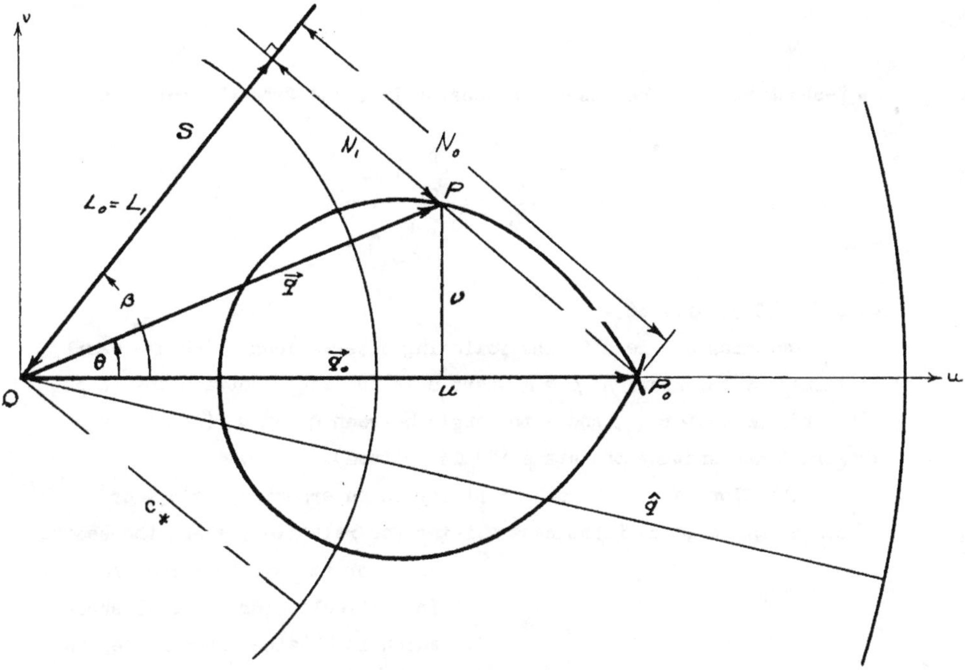

Figure 28
Construction of shock with shock polar in u,v-plane

$L_o = L$, N_o, N of the velocity vectors $\vec{q}_o$ and $\vec{q}$ with respect to the shock line can be read off from the diagram as indicated there.

The shock polar in the u,v-plane yields immediately only relations between velocities. To obtain pressure and density behind the shock front one may make use of the formula

$$p - p_o = \rho_o(N_o^2 - N_oN) = \rho_o(N_o^2 + \mu^2 L_o^2 - c_*^2), \tag{53}$$

(obtained by using (ii_o^*) of Art. 66 and (iii_p) of Art. 67), which by (49) and (50) is equivalent to

$$p - p_o = \rho_o v \tan\beta . \tag{53'}$$

Of course, the pressure could also be obtained graphically by using the

θ,p-shock polar. The change in density is found from the relation

$$\frac{\rho}{\rho_0} = \frac{\frac{p}{p_0} + \mu^2}{\mu^2 \frac{p}{p_0} + 1}$$

(see III (51), Art. 36).

Our diagrams lead to the following observations which are easily confirmed by calculation: $\beta > \theta$ where β is the angle between the shock line and the vector q_0, and θ the angle between q and q_0 (again the subscript designating the state (1) is omitted).

The flow past the shock front may be supersonic, $q > c_*$, or subsonic, $q < c_*$, the first case arising for relatively weak, the second case for relatively strong shocks. In particular for a normal shock which is obtained when P lies on the x-axis, $q = \frac{c_*^2}{q_0} = \tilde{u}$, (see equation after (50)).

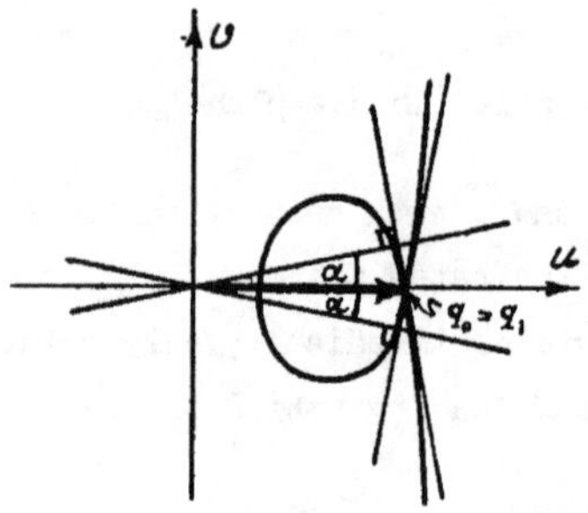

Figure 29
Shock lines tend to Mach lines for infinitely weak shocks, $q_1 \sim q_o$.

A point P with the coordinates u,v near the double point q_0 on the loop of the strophoid represents a weak shock with little change in velocity and pressure. As P tends to the double point the shock becomes sonic, the vector $\vec{q} - \vec{q_0}$ becomes tangential and hence the shock line S tends to a Mach line. Consequently the two tangents to the loop at the double point o give the normals to the Mach lines in the double point since they are orthogonal to the limiting shock line.*

The strophoid diagram shows immediately how to construct a shock from a given state (o) in front if either the shock direction β or the turning angle θ is given. It also shows that for sufficiently small angles β or θ there are two possible shocks, a weak and a strong shock transition, with small or great change in velocity (and pressure) respectively. If θ becomes small, we have in the limit either an infinitely weak sonic disturbance or a strong normal shock (see Fig. 29). There is an extreme angle $\theta = \theta_{ext}$, such that for $\theta = \theta_{ext}$ the two possible shocks coincide and for $\theta > \theta_{ext}$ no shock transition exists (see Fig. 30).

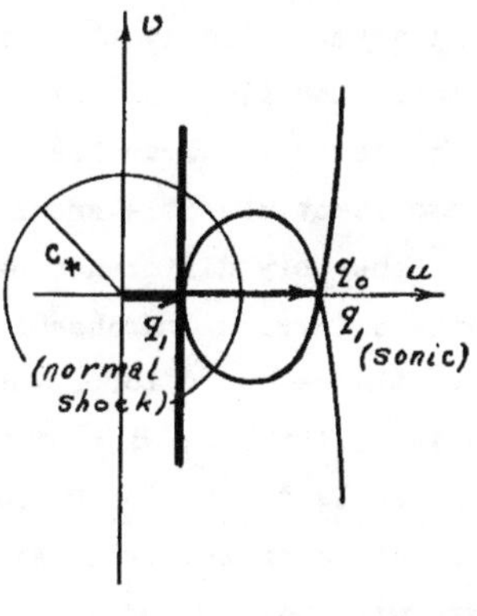

Figure 30a
Infinitely weak sonic disturbance and strong normal shock for $\theta \sim 0$.

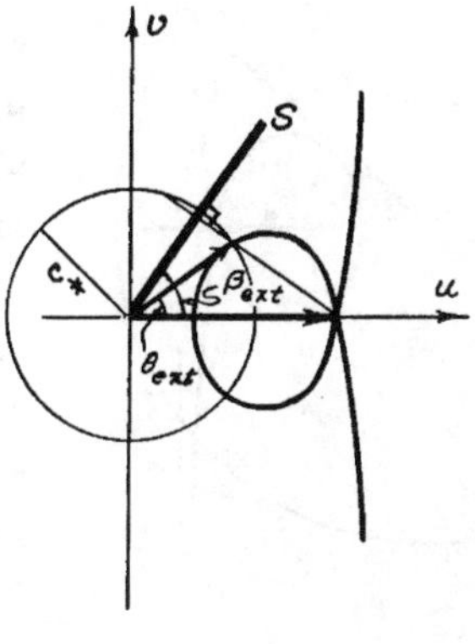

Figure 30b
Extreme angle θ_{ext} and β_{ext}.

* From this result one can easily infer the interesting mathematical result: The characteristic epicycloids of the differential equations are obtained by projecting the asymptotic lines of the pressure surface on the u,v-plane(see Art. 33).

The shock angle β reduces to the Mach angle α for a sonic shock and becomes 90° for a normal shock. In between it rises monotonically, passing through a value β_{ext} corresponding to the extreme value of the angle θ.

Incidentally, for $q_o = \hat{q}$, we have $\sin \theta_{ext} = \frac{1}{\gamma}$. In particular, for air, i.e., for $\gamma = 1.4$, we find $\theta_{ext} = 45.5$ degrees. The corresponding angle β_{ext} is $\beta_{ext} = \frac{\pi}{4} + \frac{1}{2}\theta_{ext}$ which for air is about 67.75 degrees (for this case see Fig. 33).

It should be noted that for weak shock transitions from the front (o) and for sufficiently small θ, the backward state is supersonic, while it will be subsonic for the strong shock transition. As a straightforward calculation shows, for the extreme angle θ_{ext} the backward state is subsonic.

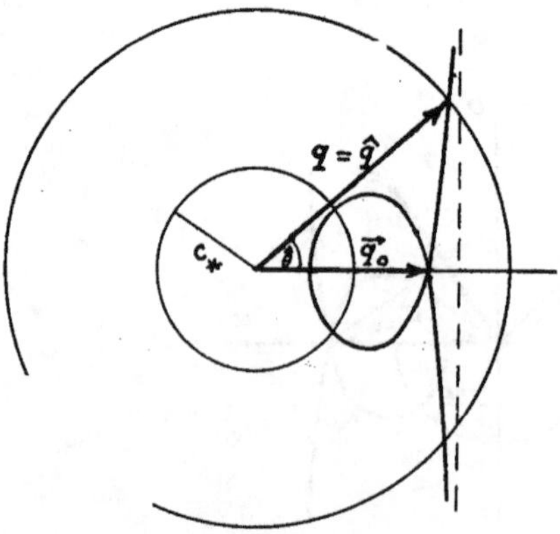

Figure 31
Limit Angle $\hat{\theta}$.

Besides the loop considered so far, the strophoid of Fig. 26 has also the two "forward branches" approaching asymptotically the line $u = U$. Within the limit circle $u^2 + v^2 = \hat{q}^2$ these two branches likewise represent possible shock transition, the only difference being that for these forward branches in contrast to the backward loop, the state (o) is on the back side of the shock front while the state on the forward branch is on the front side. Again there will be a limit angle for θ, namely, $\theta = \hat{\theta}$, the angle corresponding to the intersection of the forward branch with the limit circle.

<u>Strophoid for subsonic state (o).</u> In the preceding discussion we assumed the state (o) supersonic. Fig. 25 shows the shock polar for subsonic state (o), i.e., for $q_o < c_*$. In this case the strophoid consists of one single branch without a loop; the shock construction remains as before and, of course, always yields shock fronts for which (o) is on the back side.

For practical purposes it is convenient to have a set of strophoids with different values of q_0 prepared on transparent paper. It is then simpler to use only those strophoids with $q_0 \geq c_*$, which are sufficient to cover all cases that occur.

Limiting cases. A word should be said concerning the limiting cases which arise when q_0 approaches either c_* or $\hat{q}$. In the first case (Fig. 32) the loop shrinks to a point and the two forward branches form a cusp at this point.

If q_0 tends to the limit speed $\hat{q}$ (Fig. 33), then the forward branches approach the line $u = \hat{q}$, (which is without physical interest) while the loop approaches the circle $v^2 = (\hat{q} - u)(u - \mu^2\hat{q})$.

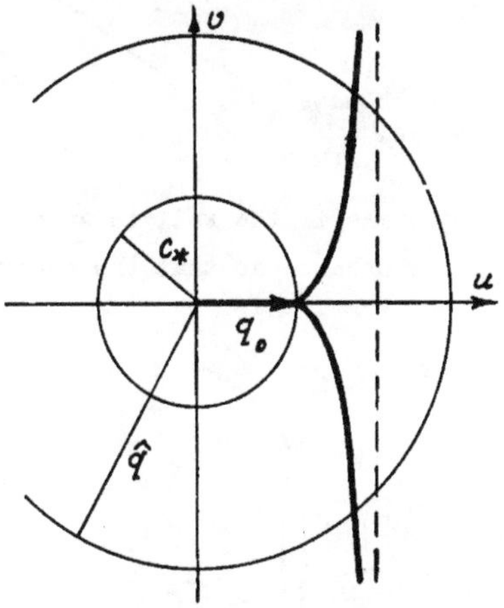

Figure 32
Shock polar for $q_0 = c_*$.

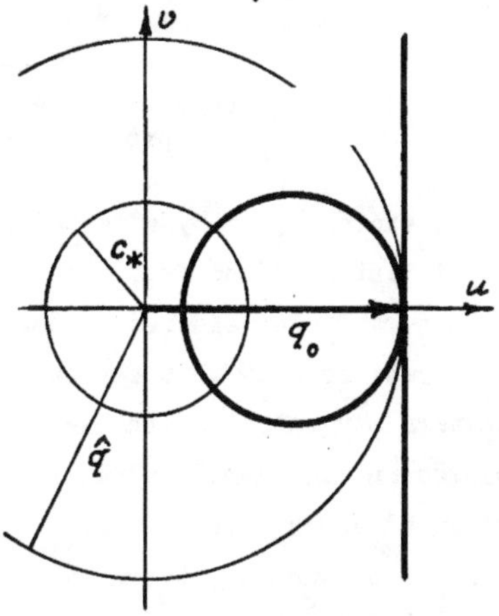

Figure 33
Shock polar for $q_0 = \hat{q}$.

C. Configurations With Several Oblique Shock Fronts. Shock Reflection.*

71. Regular reflection of a shock wave on a rigid wall. Problems of reflection occur in connection with physical situations of the following

* See the general report by von Neumann [19], and other papers quoted in the Bibliography.

type.* Suppose an approximately plane shock front moving through space hits a rigid wall obliquely. For example, such a shock wave is produced

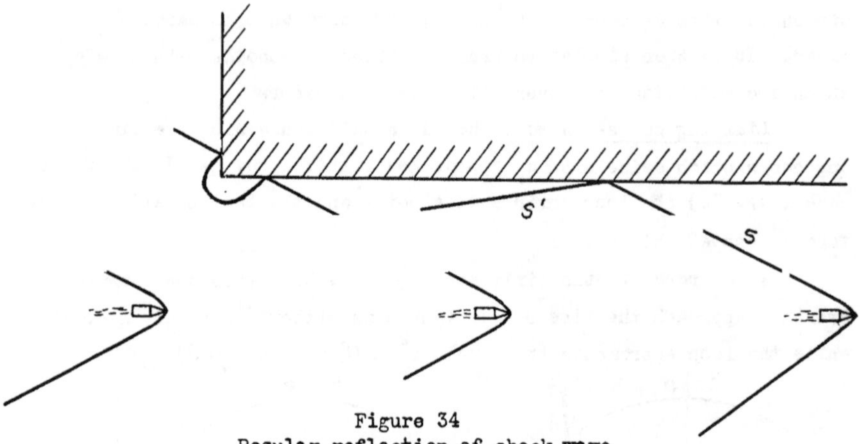

Figure 34
Regular reflection of shock wave produced by projectile.

by an explosion or by a fast-flying projectile if the wall is at a sufficient distance from the source of disturbance so that the shock fronts may be considered plane. Observations show that a flow pattern may result which exhibits an "incident" and a "reflected" shock front. (The genesis of such a pattern is indicated in the Fig. 34.)

A phenomenon similar to that of shock reflection is the interaction of two symmetric

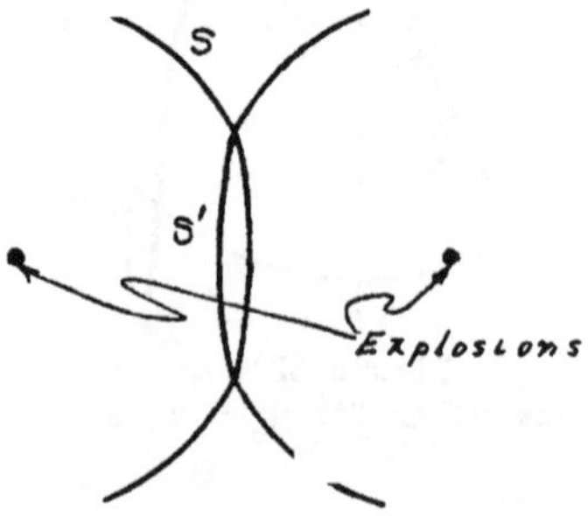

Figure 35
Regular reflection of shock wave produced by simultaneous explosion of equal charges.

* An entirely different physical situation leading to "reflection" of shocks will be discussed later in connection with flow in jets. Art. 81, Chapter V.)

shock waves meeting obliquely; the rôle of the wall will then be played by the line of symmetry of the flow. (The situation can be produced, for example, by two simultaneous explosions of equal charges, see Fig. 35.)

For very weak stationary shocks the shock lines, as seen before (Art. 70), are approximately Mach lines, and will therefore form the Mach angle with the wall, which is a streamline. Thus a stationary flow pattern containing a weak "incident" and "reflected" shock will agree with the law of reflection of geometrical optics, i.e., both shock lines will form the same angle with the wall.

There is no reason why the situation should be similar when the incident or reflected shock (or both) has appreciable strength. As a matter of fact, observations show definite deviations of the flow pattern from that of sonic waves. One speaks of regular reflection when the resulting flow resembles qualitatively that of the sonic case, i.e., when it can be described solely in terms of incident and reflected waves. Even in regular reflection the angles between incident and reflected waves are in general not equal. (The difference between the incident and reflected angle is indicated in Figs. 36, 37; one observes that the shock waves slant in the opposite direction after reflection.)

The peculiar character of the reflection of shock waves** can easily be derived and analyzed theoretically as a direct consequence of the results in Art. 70. Such a theoretical discussion consists merely in finding a mathematically possible flow pattern compatible with the observed qualitative and quantitative features of the phenomena and yielding quantitatively correct predictions of flows not yet

---

* In Art. 72 we shall discuss flow patterns which are even qualitatively quite different from sonic reflection.

** Extensive research in the theory of oblique shock reflection is due to von Neumann and his collaborators, [19], [37], [38].

observed.*

For the mathematical formulation of the problem an important preliminary step is the following. The phenomenon, as it presents itself to the observer, need not be stationary but we reduce the problem to one of steady flow by subtracting from all velocities the velocity vector (parallel to the wall) of the point O at which reflection takes place; in other words, by referring the flow to a coordinate system moving with the point O. Then we are to find a steady flow in the x,y-plane for $y \leq 0$ such that the lower half-plane $y < 0$ is divided, as in Fig. 36, into three regions (o), (1), (2),

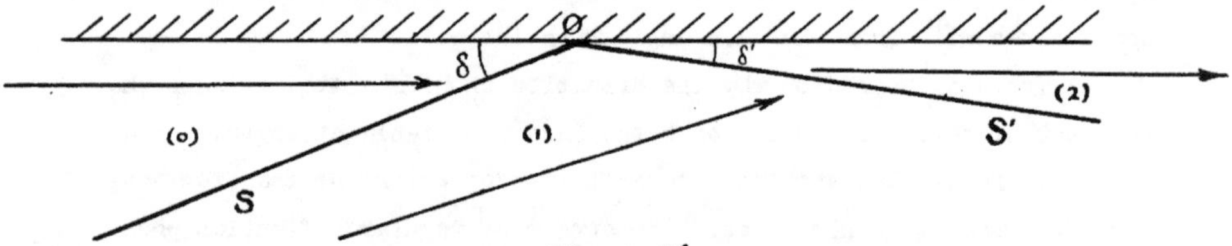

Figure 36
Regular (weak) reflection of shock fronts on a rigid wall (see Figs. 37 and 38)

each of constant state, separated by two stationary shock lines S and S' and such that in the regions (o) and (2) the flow is parallel to the wall $y = 0$, i.e., $v_o = v_2 = 0$. The configuration then consists of the succession of two oblique shocks**, the incident shock S and the reflected shock S', and it is clear from the previous sections that on passing the incident shock front the incident flow with the velocity $\vec{q_o}$ is turned toward the wall into a flow which is still supersonic with the velocity $\vec{q_1}$ and that this flow on crossing the reflected shock is turned into a flow with velocity $\vec{q_2}$ parallel to the wall, supersonic or subsonic as the case may be. (In Fig. 36 both

---

* For experimental facts we refer to the reports [37] and [39].

** One being analogous to a shock in a corner, the other to a shock outside of a corner.

shocks are facing toward the left.) From our knowledge of reflection of normal shock waves (Art. 42, Chapter III), we should expect a considerable rise in pressure behind the reflected shock front.

The mathematical objective is to find all such configurations and the corresponding interrelations between the velocities, pressures, angles, densities, etc. What the physically given or observable quantities are may vary from case to case; as long as we know all the relevant relations between q, p, ρ, c, c_*, we may base the mathematical analysis on whatever quantities are most suitable as independent variables.

Having transformed the problem into one involving steady shocks, we find it convenient to take the critical speed c_* as one of the given quantities. Regular reflection can then be discussed with the aid of the shock polars in the u,v-plane. We start from state (o) and draw the corresponding backward loop of the shock polar (strophoid). The state (1) is represented by a point on this loop. (If, for example, the shock direction between (o) and (1) is prescribed, then this point is determined as the point 1 of intersection with the loop and the normal to the shock direction through o.) Through the point 1 we again draw the backward strophoid loop (symmetric about the direction (o - 1) of $\vec{q}_1$). If, as in Fig. 37, this second loop intersects the line v = 0, then, as is easily seen, one intersection $\tilde{2}$ must be subsonic while the other, 2, may be supersonic. The states (2) and ($\tilde{2}$) are mathematically possible states behind a "reflected" shock and the reflected shock line is easily found as the line perpendicular to (1-2)

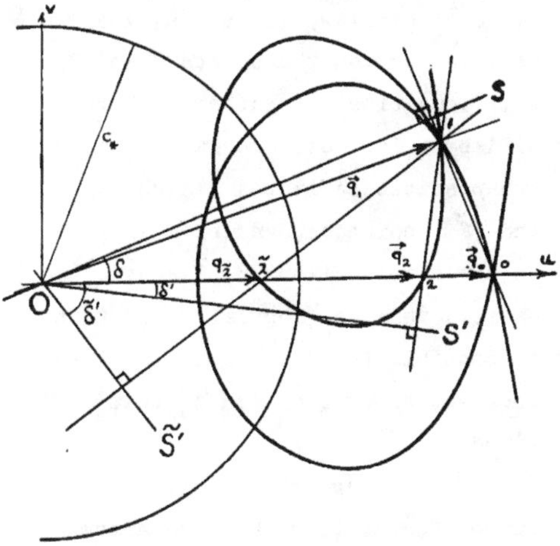

Figure 37
Construction of weak (S') and strong ($\tilde{S}$') reflected shocks (see Fig. 36).

or $(1-\tilde{2})$ respectively. Hence, there are two possibilities for a reflected shock: the regular strong reflection corresponding to $(\tilde{2})$ involving a high value of p_2, and a regular weak reflection represented by (2) and involving a smaller value of p_2. The weak reflected shock front makes a smaller angle δ' with the wall than the strong reflected shock front. Ordinarily one should expect that it is the weak reflected shock that occurs in phenomena of reflection as described above.* (The weak reflection is nearer to the sonic case, approaching it when the strength of the incident shock decreases. Moreover, it does not require an excessive rise in pressure.)

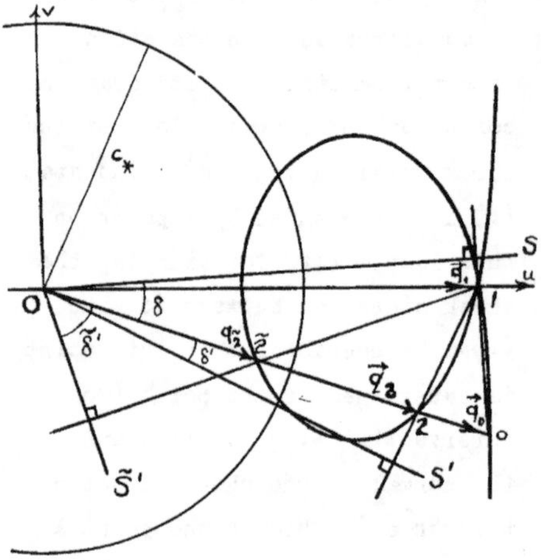

Figure 38
Alternative construction of reflected shocks (see Figs. 36 and 37).

If the point 1 on the first loop recedes from o, i.e., if the indicent shock becomes stronger or the pressure ratio $\frac{p_1}{p_o}$ greater, the second loop will shrink and the points 2 and $\tilde{2}$ will come closer together. There will be a last extreme case when the two possible reflected states (2) a and $(\tilde{2})$ coincide, and from there on the intersections 2, $\tilde{2}$ cease to exist; thus, no regular reflection is possible for sufficiently strong incident shocks, q_o and c_* being fixed.

The following is a different way of formulating the same argument. We start from state (1) and rotate the diagram about O so that the vector $\vec{q_1}$ is horizontal. Then we draw the complete strophoid through 1. The state (o) now lies on the forward branch as in Fig. 38. From the intersection of the line O-o we again

---

* See, however, Art. 81, Chapter V, where instances of strong reflection are pointed out

obtain two possible states (2) and $(\tilde{2})$, one state, or none at all, according to whether our line intersects at two points, touches, or misses the backward loop. The points (2) and $(\tilde{2})$ correspond to states behind the reflected shock. The shock directions are again immediately determined as perpendicular to the lines (o-1), (2-1), and $(\tilde{2}-1)$.

Both of these constructions yield all the relevant information about regular reflection; the quantities ρ, p, etc., are determined by the relations previously established (see also next article).

One important feature becomes evident from this construction. Regular reflection, no matter whether weak or strong, can occur only under restrictive conditions. These conditions are obtained from the diagram in a manner immediately adapted to the case of stationary shock fronts with given values of c_* and q_o. If we are interested not in steady flow, but in shock waves impinging obliquely on a zone of rest (o), the extreme situations, where regular reflection ceases to exist, must be characterized with respect to the state (o) of rest. The corresponding conditions can easily be inferred from the preceding results.

Suppose state (o) is fixed as a state of rest, and the angle δ of the impinging shock wave and the pressure ratio $\frac{p_o}{p_1}$ are considered as the two independent variables. Then c_* is a dependent variable. The relations previously obtained can then be solved numerically or graphically for c_* and q_o and thus the reflected shock can be determined by reduction to a steady state. To describe the variety of possible regular reflections we consider in a plane with the coordinates δ and $\frac{p_o}{p_1}$ the rectangle

$$0 \leq \delta \leq 90^\circ , \quad 0 \leq \frac{p_o}{p_1} \leq 1.$$

This rectangle is divided by a curve E: $\delta = \delta_{ext}\left(\frac{p_o}{p_1}\right)$, into two domains

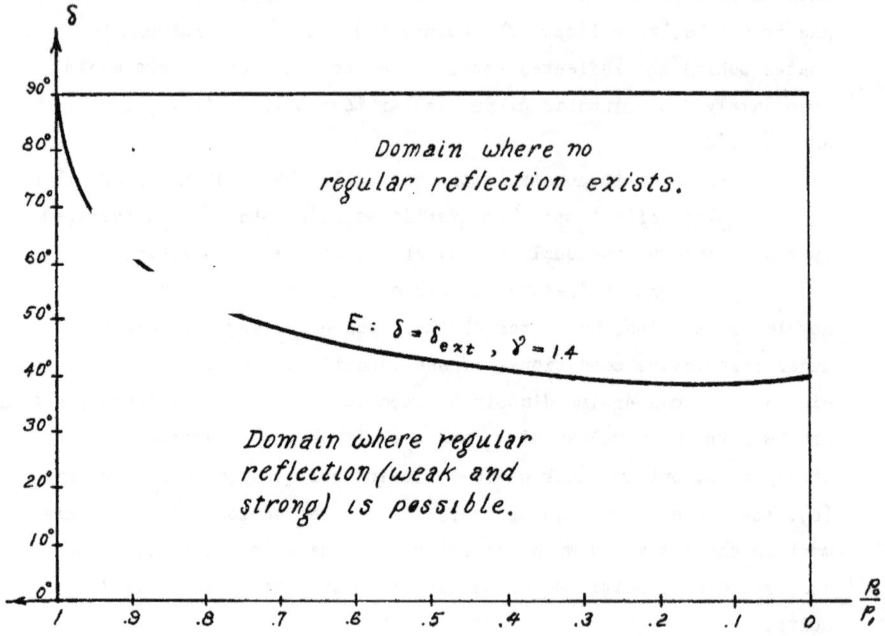

Figure 39

as in Fig. 39. To each point in the domain $0 < \delta < \delta_{ext}$ there correspond two possible regular reflections, a weak one and a strong one, as characterized above. For the data corresponding to the points in the other domain no regular reflection exists. A comprehensive picture of the manifold of regular reflection patterns is obtained by studying a sequence of incident shock waves of the same strength, i.e., having the same pressure ratio $\frac{p_1}{p_0}$, but with different angles of incidence δ. If this angle δ tends to zero then for the weak reflections the three states (o), (1), (2) tend to those occurring in a head-on reflection of a shock wave which impinges normally on the wall. Thus the reflected pressure ratio approaches for $\delta \to 0$ the limit

$$\frac{p_2}{p_1} = \frac{(2\mu^2 + 1)\frac{p_1}{p_0} - \mu^2}{\mu^2 \frac{p_1}{p_0} + 1} \tag{55}$$

given in III (68), Art. 42. As δ increases we reach the "extreme situation" represented by a point on the separating curve E. At this position the weak and the strong reflected shock coincide. As remarked above, for larger angles δ no regular reflection exists.

For the limit case of sonic shock, i.e., for $\frac{p_0}{p_1} = 1$, the extreme angle becomes 90°; for infinitely strong shocks, i.e., for $\frac{p_0}{p_1} = 0$, δ approaches the value arc $\sin\frac{1}{\gamma}$, which for air with $\gamma = 1.4$ is approximately 39.970° (this is immediately read off from the fact that the strophoid reduces to a circle for $\frac{p_0}{p_1} = 0$, see end of Art. 70).

In the memoranda [19], [38], von Neumann and Seeger have presented a full discussion with many details of regular reflection for various values of γ; they use a different approach, starting from the outset with $\frac{p_0}{p_1}$ and δ as parameters. Of the results, the information concerning the reflected pressure ratio $\frac{p_2}{p_1}$ and the angle δ' between the reflected shock and the wall is of particular interest. We give a summary for the weak reflected shock. For small values of the angle of incidence δ we have $\delta' < \delta$. As δ rises from its value in the "head-on situation", i.e., from the value $\delta = 0$, the reflected pressure ratio $\frac{p_2}{p_1}$ is below the head-on value given in (55) as long as δ' remains less than δ. Independently of the incident pressure ratio, the angles δ and δ' become equal for $\cos 2\delta = \frac{\gamma - 1}{2}$, which implies $\delta = 39.23^\circ$ for air with $\gamma = 1.4$. When $\delta = \delta'$, the reflected pressure ratio $\frac{p_2}{p_1}$ equals the value for head-on incidence. After δ has passed this value the reflected pressure ratio rises and thus exceeds the head-on value. However, only for shocks of small or medium strength $\left(\frac{p_1}{p_0} < 7.02\right.$ for air) equality of δ and δ' is attained before the extreme situation is reached. Hence for strong shocks oblique reflection will never result in reflected pressures as high as those given by (55).

These remarks about increase of pressure by reflection are obviously of practical importance; it should therefore be emphasized that for water and waterlike substances the situation is significantly different inasmuch as the rise of the reflected pressure ratio $\frac{p_2}{p_1}$ starts immediately from the head-on situation, $\delta = 0$, or δ' being greater than δ. In other words, oblique reflection in water always results in higher pressure than for head-on reflection, equal strength of the impinging shock waves being assumed.

Without relying on our strophoid diagrams one may use the shock conditions for an independent algebraic approach to the problem of reflection (and, as we shall see, to other problems concerning shock configurations). Assume that in the three domains (o), (1), (2) the thermodynamical quantities ρ and p are known (subject to the conditions (40) Art. 66, and (54), for adjacent domains). Then we have four quadratic equations (see (ii_o^*), (ii_1^*), Art. 66):

$$(56)\quad \begin{cases} \dfrac{p_1 - p_0}{\rho_0} = \vec{q}_0(\vec{q}_0 - \vec{q}_1)\,, & \dfrac{p_2 - p_1}{\rho_1} = \vec{q}_1(\vec{q}_1 - \vec{q}_2) \\[2ex] \dfrac{p_0 - p_1}{\rho_1} = \vec{q}_1(\vec{q}_1 - \vec{q}_0)\,, & \dfrac{p_1 - p_2}{\rho_2} = \vec{q}_2(\vec{q}_2 - \vec{q}_1) \end{cases}$$

from which the three velocity vectors $\vec{q}_i$ can be determined (except for an arbitrary rotation of the coordinate system) provided that one additional condition is imposed.* This condition is:

---

* The advantage is that the relations (56) do not depend on the equation of state, and thus a clear separation is possible of such features of reflection as are perfectly general, and others that are due to a specific character of the medium.

$$\vec{q_o} \times \vec{q_2} = 0$$

or $\vec{q_o}$ and $\vec{q_2}$ are parallel (in the direction of the wall). If the wall is given by $y = 0$, this means $v_o = v_2 = 0$, and we are left with the problem of determining u_o, u_1, v_1, u_2 from our four (quadratic) equations. We refrain from carrying out the analysis here.

72. Mach reflection. Experimental facts. Mathematical problem. As we have seen, regular reflection is impossible in a great many cases, in particular, when the incoming shock is very strong (for given $\frac{q_o}{c_*}$), or when a shock impinges on a quiet medium so that the angle $\beta_{o1} = \delta$ between the incoming shock and the wall is rather large.

What happens in these cases under experimental conditions similar to those which otherwise would have produced regular reflection? The answer is provided by innumerable phenomena of wave interaction* studied long ago in experiments and papers by E. Mach. Yet this "Mach reflection" as it is now called, was all but overlooked and forgotten until attention was recently drawn to it; thereafter a mathematical theory was developed by various authors with a view toward applications and understanding and control of phenomena involving considerable pressure increases occurring in oblique collision of shocks.

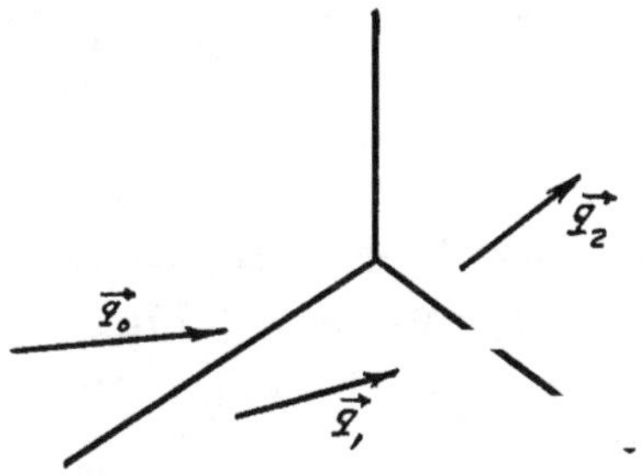

Figure 40
Hypothetical configuration of three confluent shocks.

* Observable every day in ponds and brooks and in photographs in textbooks.

The observations just mentioned show that in cases when regular shock reflection does not occur, configurations of three shocks through one point arise. It is, however, of great importance to note the following fact: Three shocks separating three zones of different continuous density and pressure are impossible. To prove this statement, we may consider a small neighborhood of the point Z where the (assumed) three shock lines meet; then for the three states near Z we have from the three shock relations

$$(57) \qquad \frac{p_i}{p_k} = \frac{\lambda\rho_k - \rho_i}{\lambda\rho_i - \rho_k}$$

with $\lambda = \mu^2 = \dfrac{\gamma - 1}{\gamma + 1}$, applied to the assumed three states (o), (1), (2). After multiplication, we find

$$(58) \quad D(\lambda) = (\lambda\rho_0 - \rho_1)(\lambda\rho_2 - \rho_0)(\lambda\rho_1 - \rho_2) - (\lambda\rho_1 - \rho_0)(\lambda\rho_0 - \rho_2)(\lambda\rho_2 - \rho_1) = 0,$$

where $D(\lambda)$ is obviously a polynomial of 2nd degree in λ and where our equation presumably is satisfied for the value $\lambda = \mu^2$, which is between 0 and 1. Likewise we see immediately $D(0) = D(-1) = 0$. Hence $D(\lambda)$, vanishing for three different values of λ, must vanish identically, and we have, for example, $D(1) = 0$, which yields

$$(\rho_0 - \rho_1)(\rho_2 - \rho_0)(\rho_1 - \rho_2) = 0.$$

Thus two of the adjacent states are identical, and our assumption of three confluent shocks is refuted.

Reflection patterns must therefore allow for an additional discontinuity, and the simplest assumption fortunately in agreement with many observations, is that of a contact discontinuity line.

The Mach reflection can be described as a configuration consisting of three shock fronts through one point and in addition a line of contact discontinuity D through the same point.* The incident shock S is followed by a reflected shock S' (S facing left in the figure). In this case, the reflection takes place, not at the wall, but at a branch point Z moving obliquely away from the wall. The point Z is connected with the wall by a perpendicular shock through which the flow is normal. Finally, there is a discontinuity line, or vortex line D from Z toward the wall.

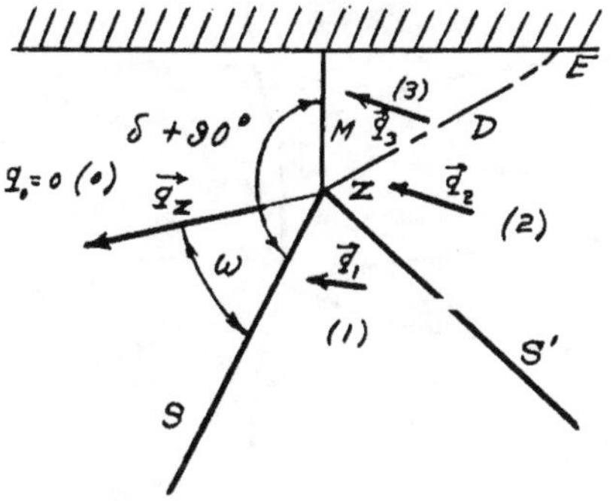

Figure 41
Direct Mach Configuration

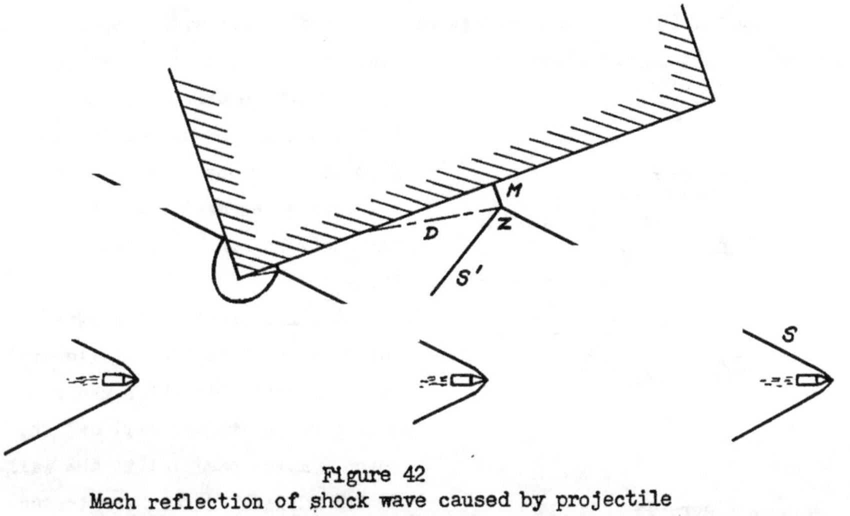

Figure 42
Mach reflection of shock wave caused by projectile

---

* As we saw, a configuration of three confluent shocks with no other discontinuity is impossible. Thus the pattern described above seems to offer itself as the scheme next in simplicity.

Flow patterns of this type are presented by nature in abundance. A possible genesis of Mach reflection as frequently observed is indicated in the Fig. 42.

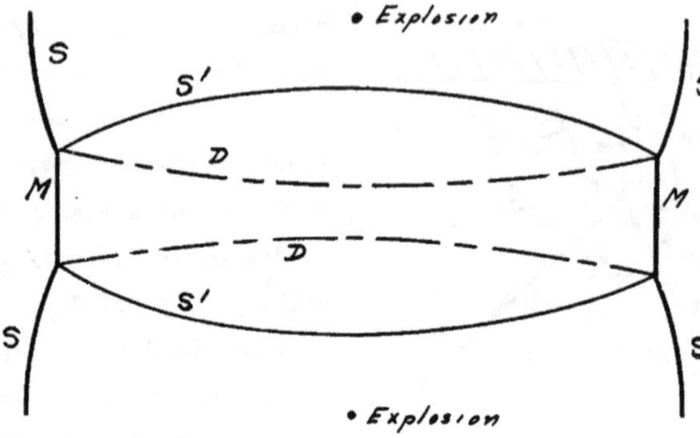

Figure 43
Mach reflection resulting from simultaneous explosion of equal charges.

Duplicating the Mach configuration by reflection with respect to the wall we obtain the flow pattern for Mach interaction of two symmetric shock fronts meeting obliquely. When two shock waves resulting from simultaneous explosion of two explosives meet, such Mach interactions can be observed (Fig. 43).

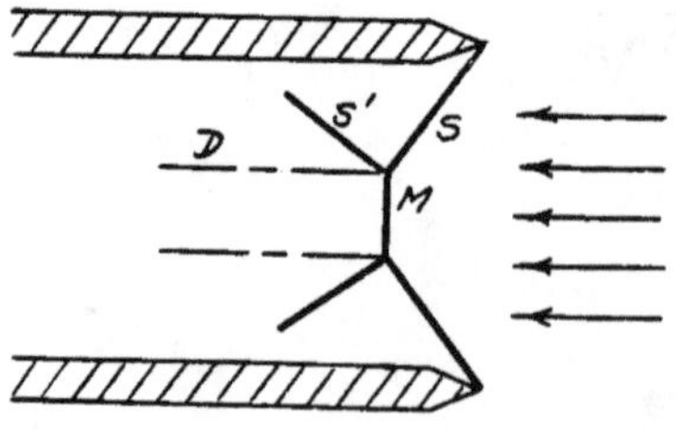

Figure 44
Stationary Mach configuration in flow against parallel wedges

A configuration of particular interest is the "stationary" Mach reflection in which the line D is parallel to the wall and the point Z moves parallel to the wall. In its duplicated form (reflected at the wall) it could occur when a steady jet of gas impinges on two parallel wedges (Fig. 44). It is to be noted, however, that the actual Mach configurations observed in jets mostly involve curved shock fronts and flow consisting of non-constant states.

It is, however, important not to be misled by the nomenclature "stationary". "Stationary Mach configuration" means simply a configuration in which the crosses the Mach shock front perpendicularly, and hence also the discontinuity line is normal to the Mach shock. Then this normal direction of the flow is automatically in agreement with the existence of a wall along which the gas flows. Many actual Mach configurations are observed in jets involving non-constant states; hence the shock lines are curved so that the stationary character of the flow and existence of a wall in no way imply that at the branch point the Mach configuration, which is characterized as a local pattern, is "stationary" in the sense defined above.

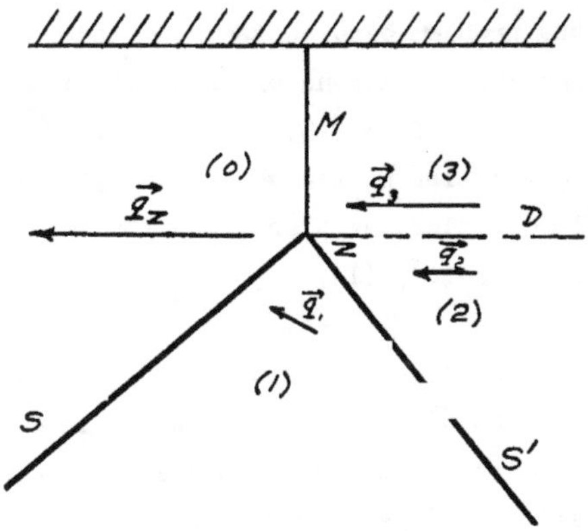

Figure 45
"Stationary" Mach reflection

In discussing the simple Mach configuration as shown in Fig. 41, it is natural to ask how the gas will flow near the point E at which the line D intersects the wall. Observed from this point, the flow along line D has the direction of D while the flow along the wall has the direction of the wall. Adjustment of these two directions will be brought about by a simple wave with center E in case the flow in (2) is supersonic when observed from E. If the flow in (2), observed from E, is subsonic a non-constant (elliptic) flow in a corner will result. In the following discussion we shall ignore this question; we shall consider only the local Mach configuration at Z assuming that the states (given by velocity, pressure and density) in each of the regions meeting at the branch point Z are constant.

To investigate all theoretically possible Mach configurations we first disregard the wall and concentrate on finding local flow patterns with three confluent shock fronts and a vortex line.

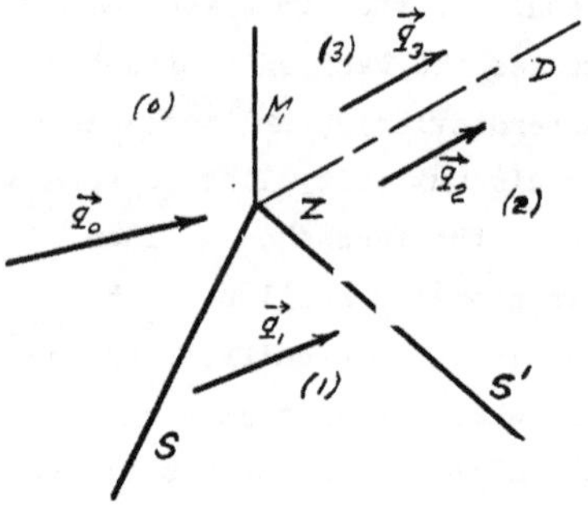

Figure 46
(Steady) direct Mach configuration.

Any three shock configuration is reduced to a steady flow by observing it from the point Z. Thus the problem becomes that of finding configurations of three stationary shock fronts S, S', M and a vortex line D through a fixed point Z, dividing the plane into four domains (o), (1), (2), (3) as indicated in Fig. 46.

After having determined velocity vectors $\vec{q}_0$, $\vec{q}_1$, $\vec{q}_2$, $\vec{q}_3$ for the four domains we need only subtract from all velocities the vector $\vec{q}_0$* in order to obtain the situation when the incident shock S impinges on a state (o) of rest and where the Mach shock M is normal (i.e., crossed perpendicularly by the flow). The velocity $\vec{q}_z = -\vec{q}_0$ is that of the branch point Z moving into the zone (o), and the discontinuity line D, of course, is given by the direction of the flow in (3). Since, as we have seen, in the steady state a shock turns the flow toward the shock line, the relative inclination of the line D and the line of motion of Z is as in the figures. After transforming (o) to rest, the flow through the shock line M will be perpendicular, and therefore a rigid wall, terminating the regions (o) and (3) is compatible with the flow. Hence, a theory of the Mach reflection essentially amounts to the construction of steady three-shock configurations with a vortex line through a fixed point.

Such a configuration may either represent an ordinary Mach reflection, namely, when the relative position of the shock M and the velocity

---

* The new vectors are again denoted by $\vec{q}$ in the diagrams.

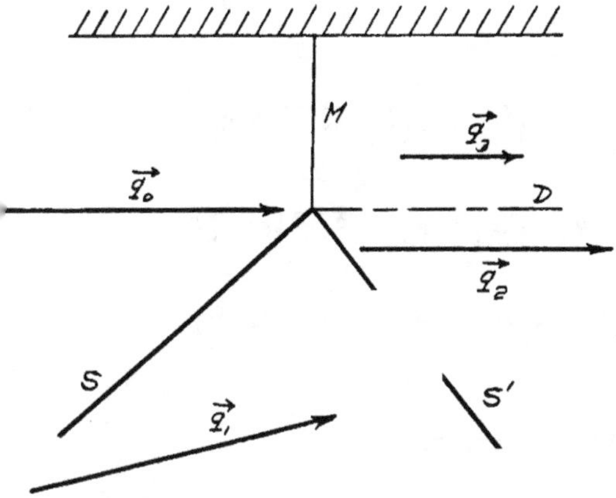

Figure 47
Stationary Mach configuration reduced to steady flow.

$\vec{q}_0$ is as in Fig. 41 (in which case the branch point Z moves away from the wall), or it may represent a so called inverted Mach reflection when the configuration is as in Fig. 47. In this configuration the point Z would move toward the wall and the configuration would quickly be destroyed. Hence this inverted Mach reflection may be eliminated from the consideration of actual reflections of shocks by rigid walls.

The transition between direct and inverted Mach reflection is the case of stationary Mach reflection, i.e., the case where the velocity $\vec{q}_0$ is perpendicular to M, or what is equivalent, when the velocities $\vec{q}_0$ and $\vec{q}_3$ are parallel. In a stationary Mach configuration the point Z will move parallel to the wall and the vortex line D is likewise parallel to the wall (see Fig. 48).

73. <u>Algebraic framework</u>. To attack the problem algebraically, one could start from the relations between pressures and velocities

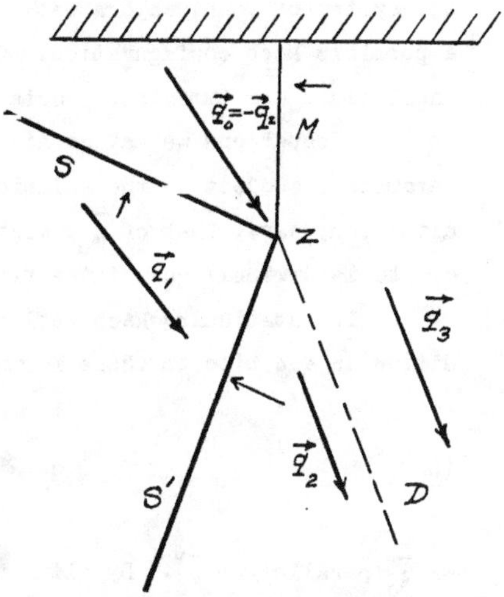

Figure 48*
Inverted Mach configuration.

* Heavy arrows denote velocities corresponding to the steady state; light arrows denote velocities corresponding to a wave motion into a quiet zone (o) (see footnote p. 214).

in adjacent domains, expressed by the six equations (see (ii_0^*), (ii_1^*), Art. 66):

$$(59)\quad \begin{cases} \tau_0(p_1 - p_0) = \vec{q_0}(\vec{q_0} - \vec{q_1}) , & \tau_1(p_0 - p_1) = \vec{q_1}(\vec{q_1} - \vec{q_o}) \\ \tau_1(p_2 - p_1) = \vec{q_1}(\vec{q_1} - \vec{q_2}) , & \tau_2(p_1 - p_2) = \vec{q_2}(\vec{q_2} - \vec{q_1}) \\ \tau_o(p_3 - p_0) = \vec{q_0}(\vec{q_0} - \vec{q_3}) , & \tau_3(p_0 - p_3) = \vec{q_3}(\vec{q_3} - \vec{q_0}) \end{cases}$$

These relations are supplemented by the condition

$$(60)\qquad \vec{q_2} \times \vec{q_3} = 0 \quad \text{or} \quad \vec{q_2} \text{ parallel to } \vec{q_3} .$$

Any system of four vectors satisfying these seven equations will give a possible Mach configuration. Once the velocities are known, the shock lines are immediately determined. As we saw previously (Art. 71), in these equations we may consider the pressures and densities as given parameters subject to the relations (40), Art. 67 or (54) Art. 70. One direction, e.g., that of $\vec{q_0}$, might be arbitrarily prescribed; then we are to determine 7 quantities from 7 equations.

The stationary Mach reflection is simply characterized by one condition in addition to those formulated above (59), (60), viz.,

$$(61)\qquad \vec{q_0} \times \vec{q_3} = 0$$

or $\vec{q_0}$ parallel to $\vec{q_3}$. By elimination of the velocities a condition involving solely densities and pressures can be obtained for stationary Mach effect. We shall not, however, follow this line of an algebraic treatment here, but rather discuss the Mach effect on a more intuitive oasis using "shock polars".

First we formulate the results of this discussion. To this end we characterize each three-shock configuration by two quantities,

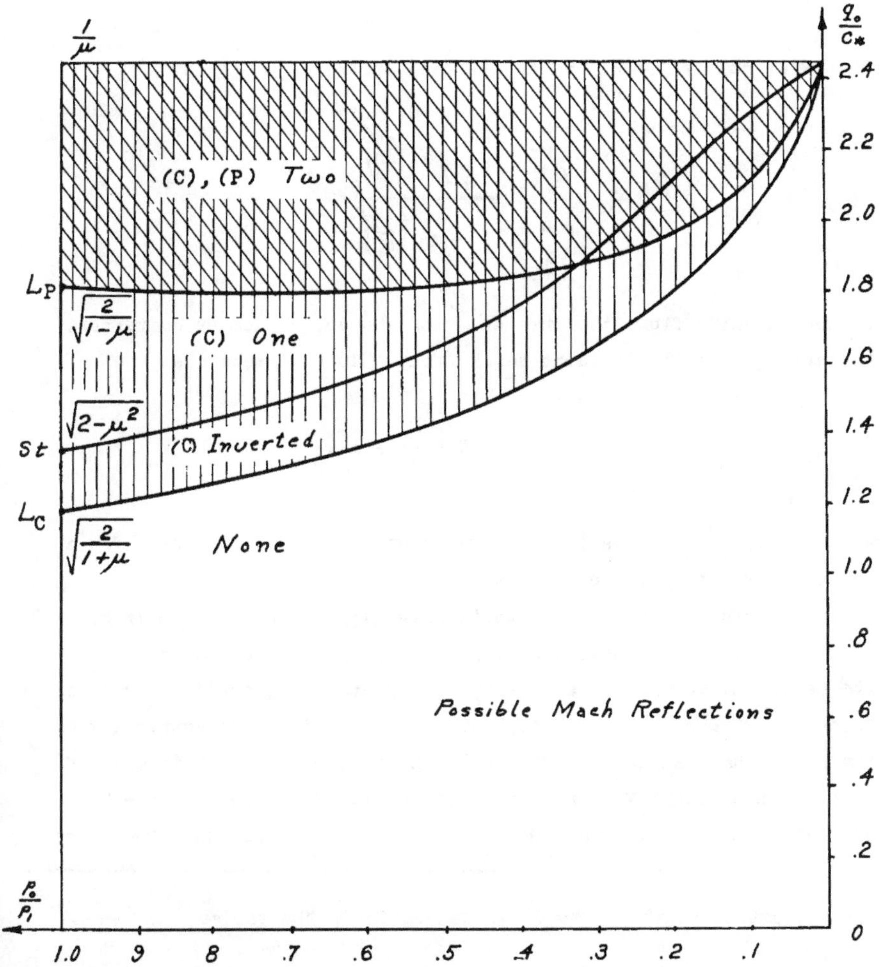

Figure 49

the numbers

$$\frac{q_o}{c_*} \quad \text{and} \quad \frac{p_o}{p_1} \ .$$

The first number is related to the Mach number $M_o = \frac{q_o}{c_o}$ by

(62)
$$M_o^2 = \left(\frac{q_o}{c_o}\right)^2 = \frac{(1-\mu^2)\left(\frac{q_o}{c_*}\right)^2}{1-\mu^2\left(\frac{q_o}{c_*}\right)^2}$$

of the incoming flow, the inverse $\frac{p_1}{p_o}$ of the second number characterize the strength of the incident shock front. In the rectangle

$$1 \leq \frac{q_o}{c_*} \leq \mu^{-1} \ , \qquad 0 \leq \frac{p_o}{p_1} \leq 1 \ ,$$

we indicate two regions (C) and (P) to each point in both of which a three-shock configuration exists.

The region (C), shaded vertically (||||||) covers the "main branch" of three shock configurations, first determined by Chandrasekhar The region is bounded by a line L_C corresponding to the limit case in which the strength of the reflected shock S' and the discontinuity of the line D have shrunk to zero while the incident shock S is aligned with the Mach shock M. The region (C) contains further a curve St corresponding to stationary Mach reflections.* The points between the

---

* The equation of the curve St is quadratic in the variable $x = \frac{q_o^2}{c_*^2}$.

With the notation $\xi = \frac{p_o}{p_1}$, the equation of St is curves L_C and St correspond to inverted, those above the line St to direct Mach configurations. The region (P), shaded obliquely (\\\\\\) bounded by a curve L_P represents an independent second branch of direct three-shock configurations, first determined by H. Polachek [38] .

$$(63)\quad \text{St:}\ \left[\xi(x-1) - \frac{1-\xi}{\xi+\mu^2}(1-\mu^2 x)\right]\left[\xi(x-1) - \frac{1-\xi}{1+\mu^2}(1-\mu^2 x)\right] = \xi(1-\mu^2 x) .$$

The angle of incidence δ is then found from

$$(64)\qquad \sin^2\delta = \frac{1+\mu^2\xi}{(1+\mu^2)\xi}\cdot\frac{1-\mu^2 x}{(1-\mu^2)x} .$$

The equations of L_C on L_p are

$$(65)\qquad x = \frac{1+\xi \pm \left[1+2\xi - (1-2\mu^2)\xi^2\right]R}{(\mu^2+\xi)\left[1 \pm (1+\xi)R\right]}$$

with

$$R = \frac{\mu}{\sqrt{(\mu^2+\xi)(1+\mu^2\xi)}}$$

the angle of incidence δ being given by

$$(66)\qquad \cos^2\delta = \frac{(\mu^2+\xi)(1-\mu^2 x)}{(1-\mu^2)\left[(\mu^2+\xi)x - (1-\xi')\right]} .$$

These formulas are equivalent to formulas (27), (30) in AMP Memo 6 and to the formulas on p. 41 in Seeger [39] .

74. Analysis by graphical methods. Substantiation and amplification of the preceding statements is appropriately based on a geometrical analysis by means of shock polars. Instead of the strophoids in the u,v-plane we use the shock polar in the θ,p-plane (52), Art. 69, which permits us immediately to take care of the condition that the

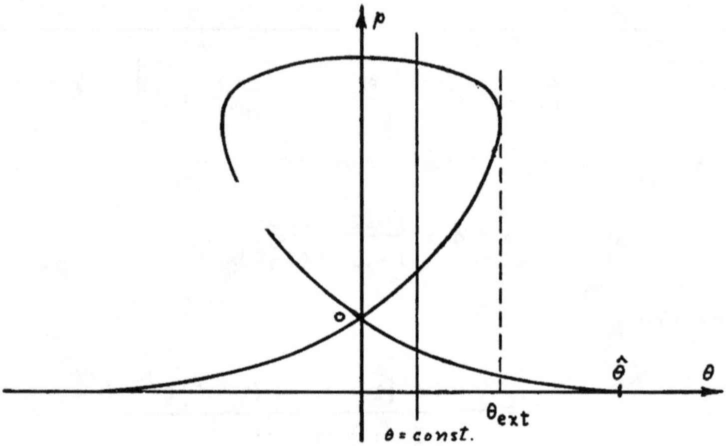

Figure 50
θ,p-shock polar, which is the image of the u,v-shock polar, showing angles θ_{ext} and $\hat{\theta}$. Loop gives possible states in back of a shock front when (o) is in front; lower branches give state in front when (o) is in back.

vectors $\vec{q_3}$ and $\vec{q_2}$, and in the stationary case, the vectors $\vec{q_o}$ and $\vec{q_3}$, be parallel. We saw that if a given state (o) is connected with another state by a stationary shock, the second state can be characterized by the angle θ through which the shock turns the flow and the pressure p in the new state. All possible states that can be connected with (o) by a shock may, therefore, be represented in a θ,p-plane by a shock polar, represented by equations*(52), Art. 69, and having the shape shown in Fig. 50, which is similar to that of a strophoid.

* Or explicitly by

$$(67) \qquad \tan\theta = \frac{\frac{p}{p_o} - 1}{\gamma M_o^2 - \frac{p}{p_o} + 1}\sqrt{\frac{(1+\mu^2)(M_o^2 - 1) - \left(\frac{p}{p_o} - 1\right)}{\frac{p}{p_o} + \mu^2}}$$

The points on the loop correspond to states behind a shock front, the state (o) at the front side $p = p_0$, $\theta = 0$, being represented by the double point o; the points on the lower branches correspond to front sides of shocks for which the state (o) is on the back side.

This polar could have been used for the general discussion of oblique shocks and of regular reflections. For instance, the extreme turning angle θ_{ext} is immediately exhibited as the maximum value of θ attained in the loop while the maximum angle $\hat{\theta}$ through which the flow can be turned if the state in back is prescribed as that corresponding to the double point is given by the point where a lower branch touches the θ-axis (attained for limit speed in front of the shock front, which means cavitation and $p = 0$). The situation resulting from regular reflection may be read off from the polar intersected by a line θ = constant as indicated in Fig. 50.

The θ,p-polar is of real advantage in the discussion of Mach configurations, which are obtainable from the diagram in such a way that numerical procedure can easily be supplied afterwards. To find Mach configurations from our shock polar Λ_0 through the point o as double point, consider a shock leading

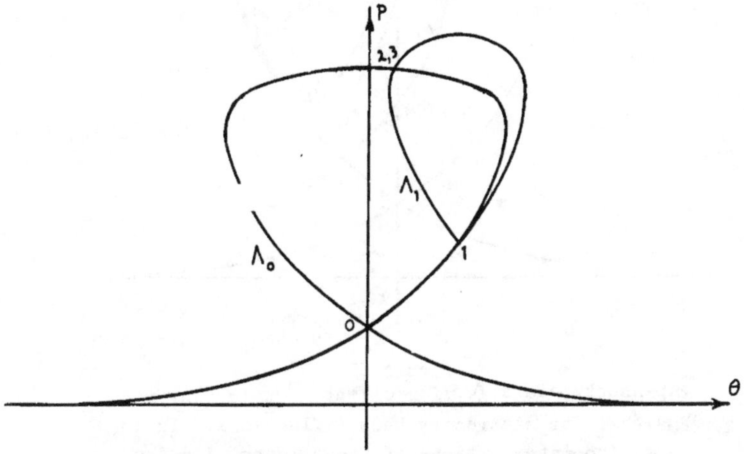

Figure 51
θ,p-shock polars for direct Mach reflection, (see Fig. 41). Point 2,3 represents states (2) and (3), on either side of discontinuity line D

from a state (o) to a state (1) with higher pressure and a shock leading from (1) to (2), as well as a shock from (1) to (3) (see Fig. 41). In the θ,p-diagram the two states (2) and (3) (see Fig. 51) on both sides of the vortex line D are represented by the same point, since in these states the pressures and the flow direction are identical. This fact is the reason for using the θ,p-diagram. We draw through the point o the shock polar loop Λ_o, and likewise another shock polar loop Λ_1 through the point 1. The point* 1 reached by a shock from (o) must be on Λ_o; likewise the point 3 is on Λ_o. On the other hand, the state (2) is reached by a shock from state (1); hence the point 2 in our diagram must lie on Λ_1, and since the points 2 and 3 in our representation are identical, we obtain 2 and 3 by intersecting the loops Λ_o and Λ_1.

If $\theta > 0$ for the point of intersection (see loop Λ_D in Fig. 52 and Fig. 41), then we have a direct or ordinary Mach reflection, and stationary Mach reflection corresponds to the case that the point of intersection occurs for $\theta = 0$, exactly at the top of the loop (see loop Λ_{St} in Fig. 52 and Fig. 48).

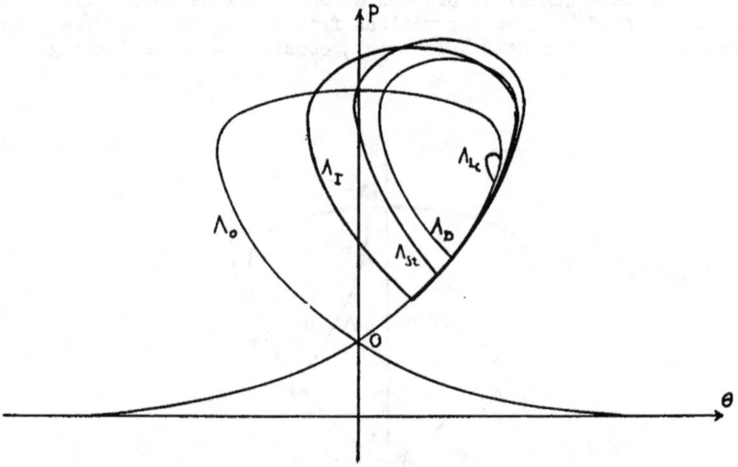

Figure 52
θ,p-shock polars Λ_D for direct, Λ_I for inverse, and Λ_{St} for stationary Mach reflections. In each case, the intersections of these loops with the loop Λ_o represent the states on either side of the discontinuity line D. (See Figs. 46, 48, 45.)

Knowing the point 2,3, we obtain the shock lines S, S' and M immediately, and then proceed to determine the densities from the relation (54), Art. 70, and the velocity vectors from our previous relations (59), Art. 73. Thereby the difference between the states (2) and (3) becomes apparent automatically since we obtain the quantities for (2) starting from (1) and those for (3) starting from (o).

Altogether, the search for Mach configurations is replaced by a discussion of our shock polar loops and their intersections. The results of such analysis may be briefly described. They confirm the diagram presented in Fig. 49.

We consider states (1) on the polar Λ_0 starting near (o), i.e., with weak incident shocks (o-1), or small $\frac{p_1}{p_0} - 1$; we then proceed to larger values of $\frac{p_1}{p_0}$. If the loop Λ_1 has an intersection with the loop Λ_0 we follow the position of this intersection until it disappears. We shall distinguish between three cases, according to the value of $\frac{q_0}{c_*}$ (which number is connected with the Mach number $\frac{q_0}{c_0}$ of the flow in region (o), (see (62), Article 73).) (Throughout this discussion reference should be made to Fig. 49. Each case in which q_0/c_* is kept constant corresponds to a horizontal line through the figure.)

<u>Case (a)</u> $\frac{q_0}{c_*}$ small, exactly: $\frac{q_0}{c_*} < \sqrt{\frac{2}{1+\mu}}$ or $\frac{q_0}{c_0} < \sqrt{2\,\frac{1+\mu}{1+2\mu}}$

Then both branches of the loop Λ_1 from 1 will start inside the loop Λ_0. Therefore no Mach configuration exists in the neighborhood (it seems that the second loop has no intersection at all with the first, so that to these values of q_0/c_* no three-shock patterns can correspond).

<u>Case (b_1)</u> $\sqrt{\frac{2}{1+\mu}} < \frac{q_0}{c_*} < \sqrt{2-\mu^2}$ or

$$\sqrt{2\,\frac{1+\mu}{1+2\mu}} < \frac{q_0}{c_0} < \sqrt{\frac{2-\mu^2}{1-\mu^2}}$$

No inverse and no stationary Mach effect exists; starting from $\frac{p_1}{p_0} - 1$ nearly zero, the intersections 2,3 always occur for positive values of θ, until for increasing p_1 a limit position L_c is reached where the second loop touches the first (tangentially) at the point 1, (see loop Λ_{L_c} in Fig. 52), the shock S' becomes sonic and the shock S is aligned with the Mach shock M, while the discontinuity along the line D disappears (see Fig. 53).

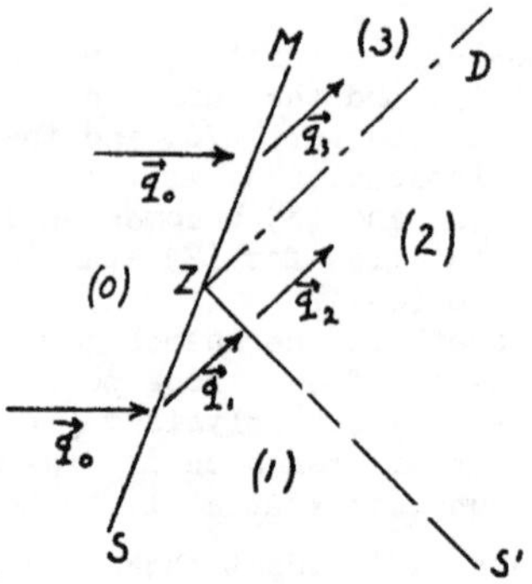

Fig. 53
Limit three-shock configuration. Incident and Mach shocks are aligned. Reflected shock S' is sonic and discontinuity of contact line D disappears.

Case (b_2)

$$\sqrt{2-\mu^2} < \frac{q_o}{c_*} < \sqrt{\frac{2}{1-\mu}} \quad \text{or}$$

$$\sqrt{\frac{2-\mu^2}{1-\mu^2}} < \frac{q_o}{c_o} < \sqrt{2\,\frac{1-\mu}{1-2\mu}}$$

(We assume $\mu < \frac{1}{\sqrt{2}}$, i.e., $\gamma < \frac{5}{3}$ from now on. For this case see Fig. 52.) (α) If the point 1 is near the point o, i.e., if $\frac{p_1}{p_o} - 1$ is small, then we have an inverse Mach effect. (See loop Λ_I).

(β) If $\frac{p_1}{p_o}$ assumes a certain value, the Mach configuration becomes stationary. (See loop Λ_{St}). (γ) If $\frac{p_1}{p_o}$ lies beyond the value for which the configuration becomes stationary we obtain direct Mach configuration (see loop Λ_D).

When the point of intersection in the θ,p-plane is vertically above (1) then the corresponding reflected shock S' will be a normal shock and from then on the shock S' will turn the flow the other way, i.e., toward the direction of the shock M. Finally, as $\frac{p_1}{p_o}$ increases we reach a limiting position L_C in which the shock S' becomes sonic as in case (b_1). (See Fig. 53 and loop Λ_{LC} of Fig. 52).

No three-shock configuration exists when $\frac{p_1}{p_o}$ increases further.

At the transition from (b_1) to (b_2), i.e., for $\frac{q_o}{c_*} = \sqrt{2-\mu^2}$, $\frac{q_o}{c_o} = \sqrt{\frac{2-\mu^2}{1-\mu^2}}$, the stationary Mach configuration is assumed for $\frac{p_1}{p_o} = 1$, i.e., when the incident shock S is reduced to a sonic wave. This situation implies the value $\delta = \arctan\sqrt{1-\mu^2}$ (= 42.4° for air, γ=1.4) for the angle which the indicent shock front S makes with the normal to the shock front M.

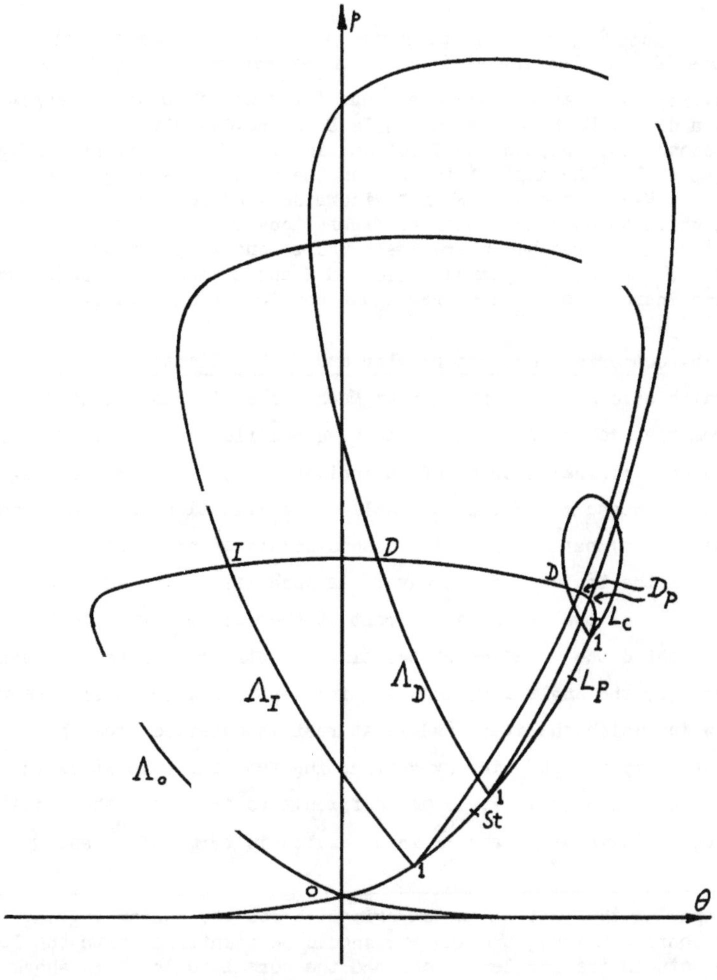

Figure 54
Several θ,p-shock polars
and their intersections.

Case (c) $\frac{q_0}{c_*} > \sqrt{\frac{2}{1-\mu}}$, $\frac{q_0}{c_0} > \sqrt{2\,\frac{1-\mu}{1-2\mu}}$

In this case the situation can be seen from Fig. 54. The loop Λ_0 through o is rather broad. For small $\frac{p_1}{p_0} - 1$ we have two configurations on loop Λ_I, one (I) indicating an inverse Mach reflection and one (D_P) indicating a direct Mach reflection. As p_1 increases, a position L_P is approached where one intersection, D still represents a direct Mach reflection while D_P coincides with (1), i.e., the second loop touches the first one in 1. For the limit configuration L_P of D_P, the angle δ is 90° and the shock S is aligned with the shock M. From now on when p_1 increases, no configuration D_P exists, while we obtain further configurations D until another limit situation L_C is reached where again the second loop touches the first. The shock S' becomes sonic and S and M become aligned. On further increase of p_1 no three-shock configuration exists.

75. Comparison between regular and Mach reflection. To discuss under which circumstances regular or Mach reflection is possible we return from the problem of a steady three-shock flow with a fixed branch point to the original problem of an incident shock wave penetrating into a zone of rest and reflected at a wall. The situation is then to be characterized by parameters which depend solely on the state (o) of rest and on the incident shock wave. As such parameters we choose (1) the ratio $\frac{p_0}{p_1}$ of the pressure in front of the incident wave to the pressure behind the wave, characterizing the strength of the incident shock and (2) the angle δ between the incident shock front and the wall.* The flow for which the state (o) is at rest is obtained from the flow with stationary branch point by subtracting from all velocities the velocity in region (o). It is not difficult to transform the results described in terms of $\frac{p_0}{p_1}$ and $\frac{q_0}{c_*}$ into results in terms of $\frac{p_0}{p_1}$ and δ.

---

* In comparing theoretical and experimental results concerning three-shock configurations, the angle δ should be identified with the local angle between the incident shock and the normal to the Mach shock at the branch point, the reason being that in reality the Mach shock is frequently curved and that therefore its direction at the branch point is in general not exactly perpendicular to the wall. (See a remark at the end of this article.)

We confine ourselves to considering direct Mach reflections belonging to the main branch. The region of points in the $\frac{p_0}{p_1}$, δ-rectangle (Fig. 55) to which such configurations correspond is shaded vertically (||||||||) This region is bounded by a curve whose points correspond to

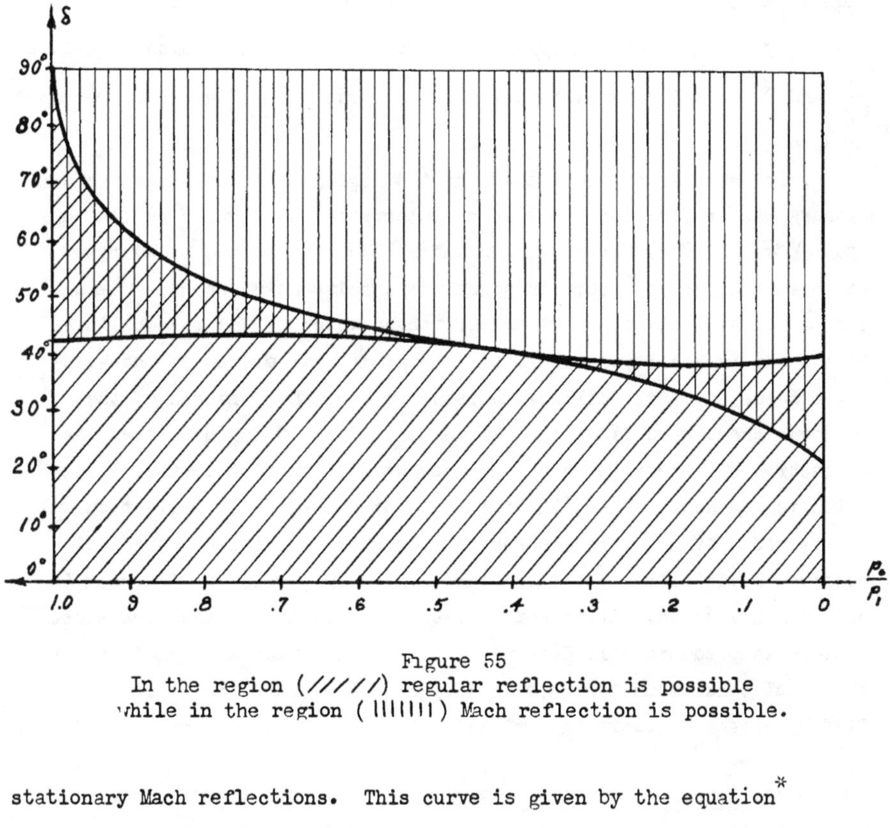

Figure 55
In the region (/////) regular reflection is possible while in the region (|||||||) Mach reflection is possible.

stationary Mach reflections. This curve is given by the equation*

$$(68) \quad \cot^4\delta - \frac{\gamma\mu^2(\xi + \mu^2) + (1 - \xi)^2}{(\xi + \mu^2)(1 + \mu^2\xi)}\cot^2\delta - \frac{\gamma(\xi + \mu^2)}{(1 + \mu^2\xi)^2} = 0$$

---

* See footnote on p. 219, (63), (64) and equations (27), (30) in AMP Memo. 6 [41].

with $\xi = \frac{p_o}{p_1}$. For $p_o = p_1$, for instance, one obtains

$$(69) \qquad \cot^4\delta - \left(\frac{1}{1-\mu^2} - 1\right)\cot^2\delta - \frac{1}{1-\mu^2} = 0,$$

whence $\tan^2\delta = 1 - \mu^2$, $\delta = 42^\circ_{,}$ for $\gamma = 1.4$, $\mu^2 = \frac{1}{6}$.

By oblique shading (//////////) we further show the region of values $\frac{p_o}{p_1}$ and δ for which regular reflection is possible. Fig. 55 corresponds to $\gamma = 1.4$.

It is clear *a priori* that the two regions overlap since the stationary Mach reflection can also be interpreted as regular reflection. (For $\gamma \leq 3.59$ (see Seeger and Polachek [38], p. 11) the boundaries touch each other at a certain point, this point being at $\frac{p_1}{p_o} = 2.61$ for $\gamma = 1.4$.) It is not clear beforehand which of the possibilities will occur in reality. According to experimental evidence, the situation seems to be as follows. Suppose we keep the strength of the incident shock wave fixed and vary δ from 0 to 90 degrees, then the reflection will be regular of the weak variety until the extreme situation is reached. Then, in discontinuous change, Mach reflection will be assumed. As δ increases further the Mach reflections as actually observed deviate from the theoretical ones, in particular for weak incident shock waves. It is not unlikely that in these cases the shock pattern is predominantly determined by the situation as a whole and that the local conditions of compatibility near the branch point are satisfied by slight unobservable twists in the directions of the shock lines.

76. Pressure relations. For Mach reflection as well as for regular reflection much of the interest is focused on the resulting pressures. Let us consider the ratio $\frac{p_2}{p_1}$ of the pressure behind and in front of the reflected shock front S'. Then the "reflected pressure ratio" varies when the angle δ of incidence varies from $\delta = 0^\circ$ to $\delta = 90^\circ$ while the strength of the incident shock $\frac{p_1 - p_o}{p_o}$ is kept fixed.

As remarked in Art. 71 (see (55)), for head-on collision, $\delta = 0$, we have

$$\frac{p_2}{p_1} = \frac{(2\mu^2 + 1)\frac{p_1}{p_0} - \mu^2}{\mu^2 \frac{p_1}{p_0} + 1} \quad (70)$$

When δ increases, the ratio $\frac{p_2}{p_1}$ will first decrease, but when δ approaches the extreme value, it may again increase beyond the head-on valuo (see remarks on p. 207). After a Mach configuration has appeared in the flow, this rise will eventually terminate and the ratio $\frac{p_2}{p_1}$ will decline again, approaching the value 1 as δ approaches 90°, since in this limit case the reflected shock becomes sonic.

77. Modifications and generalizations. The three-shock configuration studied above is a simple mathematical pattern into which many phenomena fit surprisingly well. There exists, however, much experimental material that is clearly in disagreement with theoretically determined Mach configurations. One possibility which would explain such deviations was mentioned above, viz., the actual flow pattern may correspond to a theoretical three-shock configuration only locally in an unobservably small region. There is, however, another possibility to explain such deviations, viz., by modifications of the simple Mach pattern in the hope of constructing a

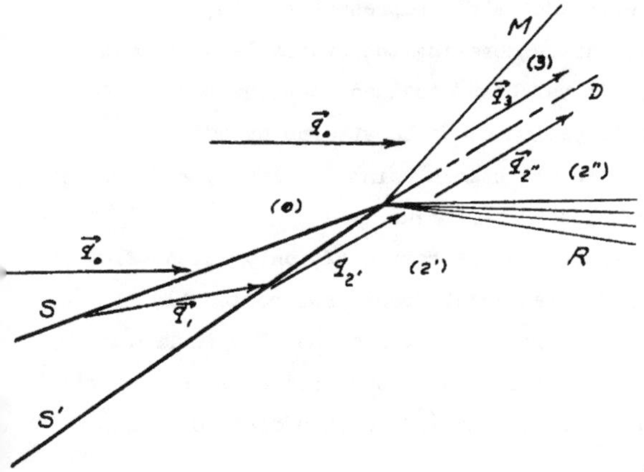

Figure 56
Three-shock configuration modified by rarefaction wave

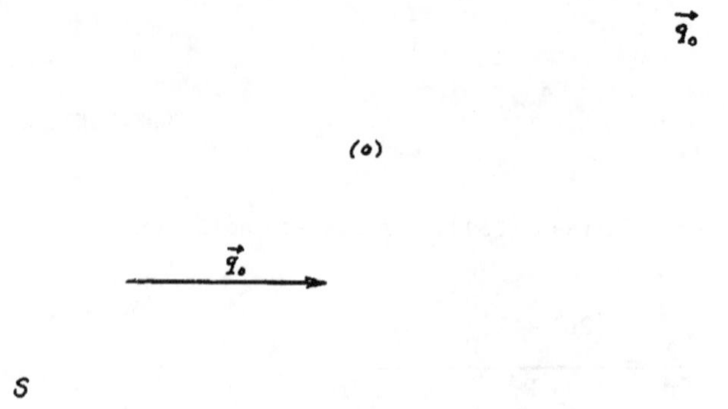

Figure 57
Flow against an "arrowhead"
(see Fig. 56)

mathematical scheme which will adequately represent reality.

Complete success, as far as covering the available experimental evidence is concerned, has not yet been attained, but the most direct generalization, namely, modification of Mach patterns by additional simple waves with the center in the branch point is already sufficient to explain some of the more elusive phenomena.

Such waves can be inserted in the region (2) only if the flow direction on crossing the reflected shock front from region (1) to region (2) is turned toward the branch point and the flow in region (2) (when observed from the branch point) is supersonic (more complete conditions are given below). The region (2) is then divided into

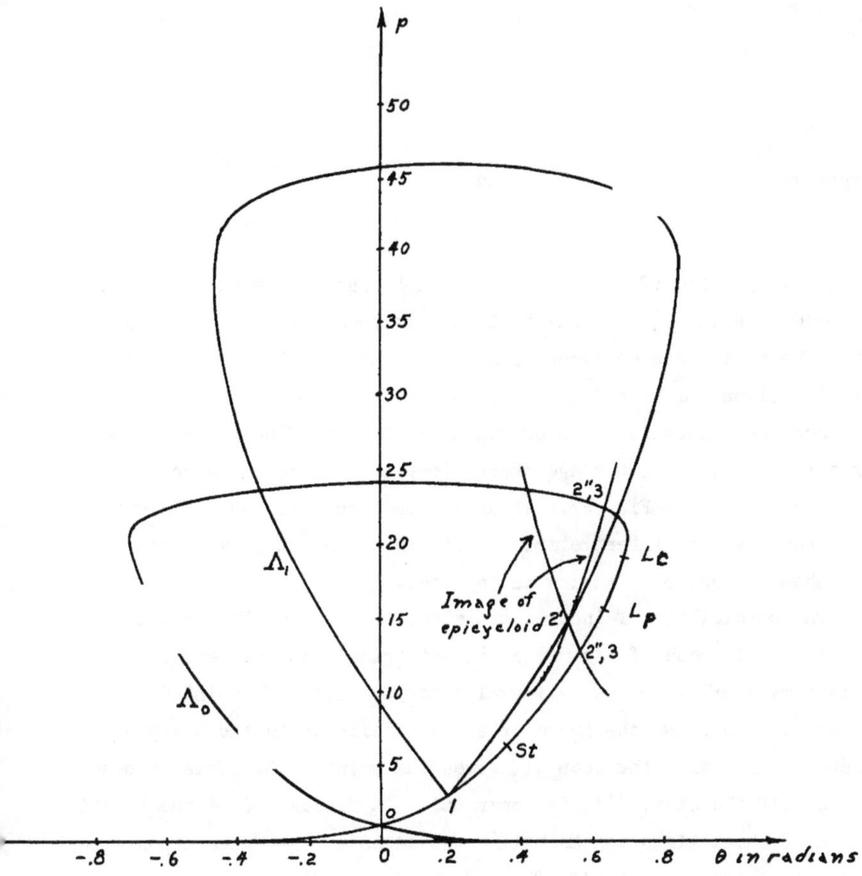

Figure 58
Representation of modified Mach
configuration in θ,p-plane (see Figs. 56, 57)

CONFIDENTIAL

two regions (2') and (2'') and the flow on crossing from (2') through the simple wave into (2'') will be turned further toward the branch point. The conditions mentioned can be seen to be satisfied if the angle of incidence δ is sufficiently near to 90°.

Such configurations can be expected when the flow is deflected by an arrow-head, i.e., a wedge whose slope after some distance changes abruptly, cf. Fig. 57. Actually observed "arrow-head" flows seem to confirm this interpretation (note that the simple wave would hardly show a trace on a shadow photograph).

The possibility of inserting a simple wave under the conditions indicated can be read off from the θ,p-diagrams discussed earlier. The diagrams in Fig. 58 are self-explanatory. Through a point 2' on the loop Λ_1 one draws the image of a characteristic in the u,v-plane; its intersection with the loop Λ_0 gives the point 2",3. This is possible only if the state (2') is supersonic. A discussion of the relative position of shock lines and Mach lines would show* that then a possible flow configuration results if $\theta_0 < \theta_1 < \theta_{2'} < \theta_{2''}$.

---

* See AMP Memo No. 38.9M [42].

V. FLOW IN THREE DIMENSIONS

78. Introduction. Flow in one dimension and steady isentropic flow in two dimensions could be treated with a fair degree of completeness because of the special character of the underlying differential equations and because of the existence of simple waves. Flow in three dimensions, however, even under restricting assumptions of symmetry which make a reduction to two independent variables possible, presents a far more complicated type of mathematical problem. It may be that extensive numerical computations of special examples could produce valuable clues for a more general theoretical attack. As matters stand now, however, one has to be content with a theoretical analysis of some particularly simple types of problems.

In this chapter we shall consider three distinct topics: axially symmetric flow through nozzles and jets, flow against a conical obstacle, and flow with spherical (or cylindrical) symmetry. The first two flows are essentially steady in character while the third is not.

A. Flow in Nozzles and Jets

79. Nozzle flow. The first topic is amenable to a simple but satisfactory approximate analysis. Strictly speaking, flow through nozzles and jets should be considered as a steady, isentropic, irrotational flow with symmetry about the x-axis. An approximate treatment of great practical value, however, is possible without entering into difficult manipulations with the differential equations.

Some qualitative features of flow in two-dimensional jets have already been discussed in Art. 64, Chapter IV. Here we shall deal with the more realistic and highly important problem of the flow through a "de Laval Nozzle" with symmetry about the x-axis. Such nozzles play a decisive rôle in the operation of turbines, wind tunnels, and rockets. The de Laval nozzle consists of a converging "entry section" and a diverging "exhaust section". When a gas at rest in a container or "chamber" under high pressure escapes through such a nozzle, two possibilities arise. The first is that the flow is expanded in the entry section, in which case the flow remains subsonic throughout. This occurs when the ratio of chamber pressure to the pressure outside remains below a certain "critical" value. When this pressure ratio exceeds the critical value, the other alternative occurs; the flow becomes supersonic on passing the throat and is expanded from there on.

80. Flow through cones. The important fact that subsonic flow is contracted, supersonic flow is expanded in a diverging section can best be recognized by considering flow in a cone. We assume that the flow is steady and isentropic, that it is radially directed, that speed q, density ρ and pressure p depend only on the distance r from the tip. Then the continuity equation can be written

$$(r^2 \rho q)_r = 0 \ ,$$

or $r^2 \rho q = \text{const.}$ Denoting the area intercepted by the cone on the sphere $r = \text{const.}$ by $A = A_0 r^2$ and the rate of mass crossing this area per unit time by G we have

$$A \rho q = G \ . \tag{1}$$

Since the flow is automatically irrotational, the only additional equations needed to determine the flow are the adiabatic relation

$$p \rho^{-\gamma} = \text{const.}, \tag{2}$$

and Bernoulli's law

$$\mu^2 q^2 + (1 - \mu^2) c^2 = c_*^2 \tag{3}$$

with $c^2 = \gamma p/\rho$.

The critical speed $c_* = q_*$ is, in what follows, always considered as a fixed parameter; then, by (1), (2) and (3), the critical values p_*, ρ_* of pressure and density are also fixed, and likewise the critical value A_* of the cross-section A, corresponding to the value $c = c_*$. (These critical values are well-defined quantities, whether or not they are attained in the actual flow.) The preceding equations may now be written in the following forms, in which c/c_*, ρ/ρ_*, p/p_*, A/A_* are expressed in terms of $q/q_* = q/c_*$:

$$(4) \qquad \left(\frac{c}{c_*}\right)^2 = \frac{1-\mu^2\left(\frac{q}{q_*}\right)^2}{1-\mu^2} \qquad (q_* = c_*),$$

$$(5) \qquad \frac{\rho}{\rho_*} = \left(\frac{c}{c_*}\right)^{\frac{2}{\gamma-1}},$$

$$(6) \qquad \frac{p}{p_*} = \left(\frac{c}{c_*}\right)^{\frac{2\gamma}{\gamma-1}}$$

$$(7) \qquad \frac{A}{A_*} = \left(\frac{1-\mu^2\left(\frac{q}{q_*}\right)^2}{1-\mu^2}\right)^{-\frac{1}{\gamma-1}} \left(\frac{q}{q_*}\right)^{-1}.$$

Accordingly, we may consider any one of the quantities A, q, r as independent variables (always for fixed $c_* = q_*$, p_*, ρ_* and A_*) and express all the other quantities in terms of it.

For a further discussion we derive from (1), (2), (3) the relations

(8) $$\frac{dA}{A} + \frac{d\rho}{\rho} + \frac{dq}{q} = 0 ,$$

(9) $$\frac{d\rho}{\rho} = \frac{2}{\gamma - 1}\frac{dc}{c} = \frac{1-\mu^2}{\mu^2}\frac{dc}{c} , \text{ and}$$

(10) $$\mu^2 q dq + (1-\mu^2) c dc = 0; \text{ hence}$$

$$\frac{dA}{A} = \left(\frac{q^2}{c^2} - 1\right)\frac{dq}{q} .$$

The last relation shows that for increasing area A the speed q increases when $q > c$ and decreases when $q < c$. Moreover, since increasing speed corresponds to decreasing density, it follows that in the direction of increasing area A the flow is expanded when it is supersonic, contracted when it is subsonic.

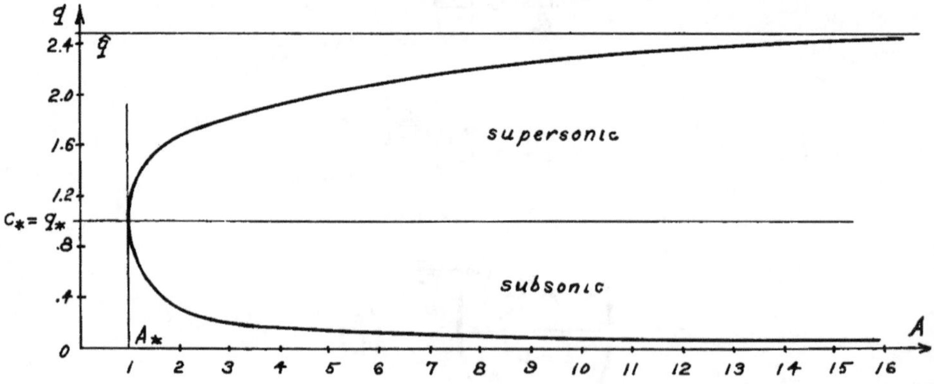

Figure 1
Speed q of flow depending on intercepted area A, (plotted for air, $\gamma = 1.4$).

Another important consequence can be inferred from these formulas. The quantity A as a function of q has a minimum A_*

for $q = q_* = c_*$. No flow with given c_*, p_*, A_* is possible in the part of the cone with $A < A_*$, and a converging flow beginning with $A > A_*$ stops when the critical area $A = A_*$ is reached. No transition from subsonic to supersonic flow is possible in a cone.

81. De Laval's Nozzle.* Such transition becomes possible, however, by the following modification. Two sections of cones or similarly shaped tubes with the same axis are placed opposite to each other and connected as in Fig. 2; thus forming a de Laval nozzle with entry section, throat, and exhaust section. Then a

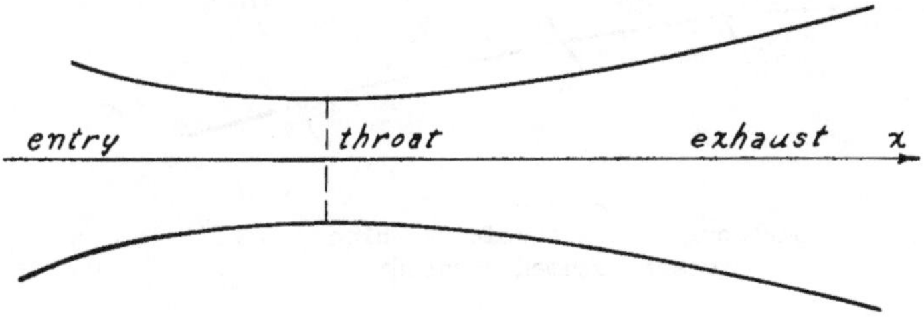

Figure 2
De Laval nozzle.

subsonic expanding flow in the entry section, on passing through the throat, can turn into a supersonic expanding flow in the exhaust section.

The preceding exact treatment of the conical flow can now be used to describe approximately the flow through such a de Laval nozzle, whose components are not necessarily conical. The approximate treatment described below is a slight modification of the "hydraulic" treatment due to O. Reynolds (1886). A set of surfaces of revolution is introduced which intersect the nozzle wall perpendicularly;** it is then assumed that the flow is orthogonal to

* With reference to what follows see [44] to [50] in the Bibliography.

**In Reynolds' "hydraulic" treatment planes perpendicular to the axis are assumed instead of curved surfaces, which are suggested when one starts with the consideration of conical flow.

these surfaces and that all relevant quantities are constant on them.

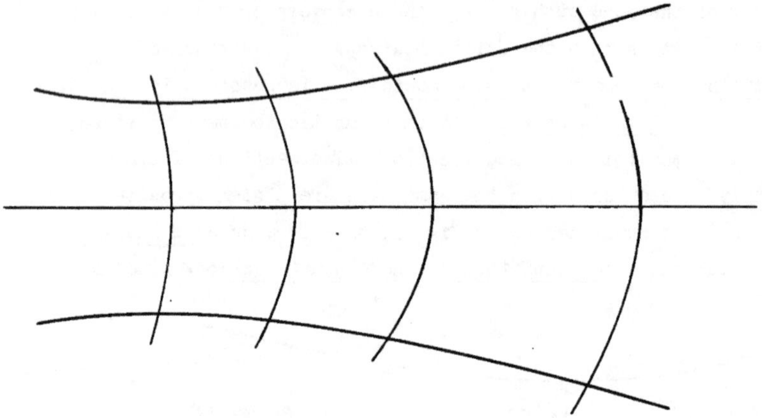

Figure 3
Surfaces across nozzle on which
q, p, ρ are assumed constant.

Denoting by A the area on these surfaces intercepted by the nozzle wall, the cross-section area, we apply the same formulas derived above for the conical flow. In particular, we infer from relation (11) that if the flow changes from a subsonic to a supersonic state at all, it becomes sonic at the throat, i.e., $q = q_* = c_*$, at that surface on which the area A assumes its minimum $A = A_t = A_*$. The assumptions on which this treatment is based are not exactly compatible with the irrotational character of the flow. Nevertheless, both experiment and a refined theoretical treatment* based on a more complete analysis of the differential equations show that the results of the hydraulic theory provide very good approximations.

It might seem that an exact theory of two-dimensional nozzle flow, (see Art.64, Chapter IV) could be developed by investigating the linear differential equations characterizing the flow in the hodograph plane. Such a procedure, however, is not possible. To see this, one need

* See AMP Report 82.1R [49].

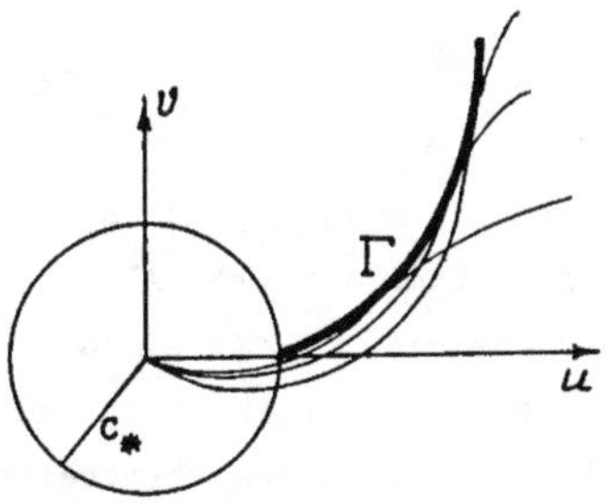

Figure 4
Images of streamlines of nozzle flow (for $y > 0$) in the hodograph plane with envelope, which is a characteristic Γ.

only imagine how the images of the streamlines would run in the hodograph plane. Consider a point of a streamline in the subsonic region with negative velocity component v. On the same streamline v will eventually be positive in the supersonic region. Since $v = 0$ all along the axis of the nozzle, one arrives at the diagram shown in Fig. 4, corresponding to the section of the flow with $y > 0$. One observes that the images of the streamlines intersect each other. Thus the image of the field of flow in the hodograph plane is not simple and the functions representing x,y in their dependence on u,v become singular at the envelope of the images of the streamlines. This envelope is necessarily a characteristic Γ, an epicycloid, because along the envelope the Jacobian $x_u y_v - x_v y_u$ becomes infinite and, just as jump discontinuities can occur only along characteristics (see Art. 11, Chapter II), it can be shown that the line along which derivatives become infinite is also a characteristic. Evidently it is the epicycloid with the cusp at the point $(c_*,0)$. Thus a treatment of nozzle flow in the hodograph plane is impossible.

82. Various types of nozzle flow. The hydraulic nozzle theory, in spite of its great simplicity, accounts for the various peculiar types of nozzle flow that result under various conditions. For a proper understanding of these occurrences it is convenient to assume that the exhaust flow is discharged into a large receiver vessel in which an arbitrary pressure can be maintained. We then imagine that the receiver pressure p_r is varied while the chamber pressure p_c at the entrance of the flow into the nozzle is kept fixed. Then the critical pressure p_* is determined by

$$p_* = (1 - \mu^2)^{\frac{\gamma}{\gamma - 1}} p_c , \tag{12}$$

assuming that the flow speed q_c in the chamber equals zero (which corresponds to infinite cross-section area of the chamber). (Formula (12) follows from $\frac{p_*}{p_c} = \left(\frac{c_*}{c_c}\right)^{\frac{2}{\gamma-1}}$ by (6), where the sound speed c_c in the chamber is given by $(1 - \mu^2)c_c^2 = c_*^2$ according to (3) and $q_c = 0$.)

Let p be the pressure at the "cross-section" of area A, then by formulas (7), (4) and (6) of the preceding article the ratio A/A_* is a well defined function of the ratio p/p_*; p_* and A_* being critical pressure and cross-section area, respectively. While the critical p_* is known, the critical area A_* is not determined by the given state in the chamber.

To visualize the variety of flows compatible with a fixed state in the chamber and various receiver pressures, it is advantageous to

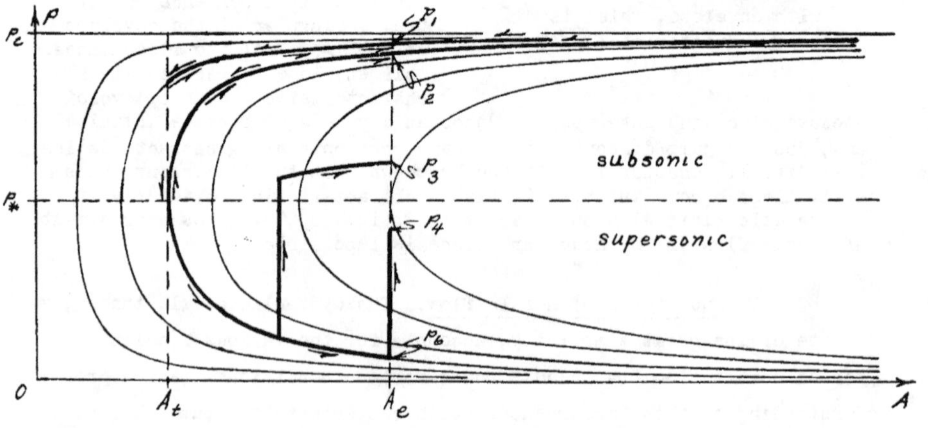

Figure 5
Relation between pressure and cross-section area for the various nozzle flows resulting from various receiver pressures.

show graphs of the set of functions

$$A = A_* f\left(\frac{p}{p_*}\right)$$

for various values of the parameter A_*, derived from (7), (4) and (6). (See Fig. 5)*

---

* Of course, this figure is closely related to Fig. 1.

All these curves have the lines $p = 0$ and $p = p_c$ as asymptotes and are loops reaching from $A = A_*$ to $A = \infty$. For any of these curves there are two values p attached to every $A > A_*$, the greater value of p referring to a subsonic, the smaller to a supersonic state, while for $A = A_*$ both states become identical, and for $A < A_*$ no flow is possible at all. In Fig. 6 the pressure p is shown as function of the abscissa x along the axis resulting when for a given nozzle the value A as function of x is inserted in $A = A_* f\left(\frac{p}{p_*}\right)$. The relation between pressures and areas along any flow in the nozzle from chamber to receiver is represented by arcs of these graphs.

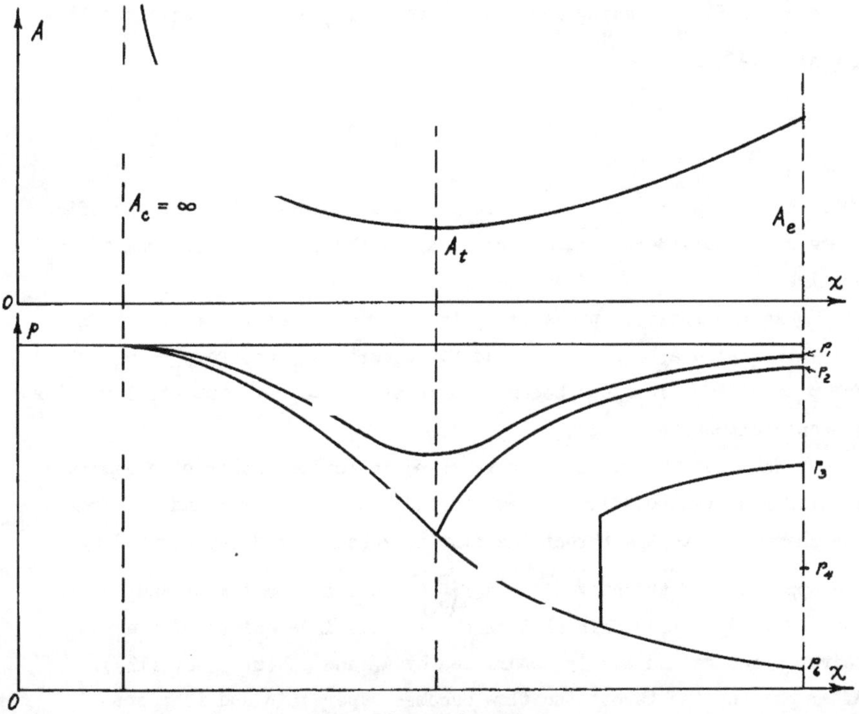

Figure 6
Pressure as a function of the position along the axis of a nozzle for various flows resulting from various receiver pressures.

When the receiver pressure p_r equals the container pressure p_c no flow results at all. When p_r is slightly less than p_c, a flow with low speed results (1). To determine it, one locates in the A,p-plane the point $p = p_r$, $A = A_e$, A_e being the exit cross section area. For an appropriate value of A_*, $A_* = A_*(p_r, A_e)$, the curve $A = A_* f\left(\frac{p}{p_*}\right)$ passes through this point: we follow this curve until A assumes the value A_t of the throat cross-section area. The section of the curve between A_t and A_e represents the flow from the throat to the exit, while the flow from the container to the throat is represented by the section from $A = \infty$ to $A = A_t$ of the same curve. The flow remains subsonic throughout. Evidently this description of the flow, characterized by the subscript $(_1)$, is valid only if the curve $A = A_* f\left(\frac{p}{p_*}\right)$ passing through the point (p_r, A_e) intersects the line $A = A_t$; i.e., if

$$A_t > A_*$$

where $A_* = A_*(p_r, A_e)$ is the critical area associated with the specific curve under consideration, and where A_t is the fixed throat area of the nozzle.

As the receiver pressure p_r is lowered, A_* decreases until finally the value $A_t = A_*$ is reached for a certain pressure $p_r = p_2$. For $p_r = p_2$ the flow just becomes sonic at the throat, but still remains subsonic elsewhere.

When now the receiver pressure p_r is lowered below p_2 a completely different type of flow arises, as indicated in Fig. 5 and 6. From the chamber up to the throat the flow is subsonic and represented by the upper arc of the curve, $A = A_t f\left(\frac{p}{p_*}\right)$, coming from $A = \infty$ and belonging to $A_* = A_t$. This part of the flow is indeoendent of the receiver pressure and solely determined by A_t and p_* (or p_c by (12)). After passing the throat the flow becomes supersonic and is represented by the lower branch of the same curve $A = A_t f\left(\frac{p}{p_*}\right)$; this curve intersects the line $A = A_e$ at a definite point with $p = p_6$; hence p_6 is determined from $A_e = A_t f\left(\frac{p_r}{p_*}\right)$. In other words, the flow (6) is a smooth flow with steadily increasing speed, sonic at the throat, and steadily decreasing pressure and density. If the receiver pressure happens to be exactly $p_r = p_6$ we thus have what is considered the ideal flow through the nozzle. The design of a nozzle for given pressures p_c and p_r usually means the selection of the dimensions A_t and A_e so that, in our notation, $p_6 = p$; i.e., $\frac{A_e}{A_t} = f\left(\frac{p_r}{p_*}\right)$

In our imagined experiment, however, when p_r is gradually lowered, we still have $p_r > p_6$, after first passing $p_r = p_2$. How does the flow after having attained supersonic speed behind the throat, adjust itselt to the prescribed receiver pressure p_r? The answer is that first the flow continues behind the throat as indicated by the lower branch of the specific curve $A = A_t f\left(\frac{p}{p_*}\right)$; but at a certain place in the diverging part of the nozzle a shock front interferes, the gas is compressed and slowed down to subsonic speed. From there on the gas is further compressed and slowed down; the relation between pressure and area is then represented by the upper branch of the curve $A = A_* f\left(\frac{p}{p_*}\right)$ passing through A_e and p_r with an appropriate smaller value of p_*. The position and strength of the shock front is automatically adjusted so that the end pressure at the exit becomes p_r. In the diagram the place corresponding to the shock front indicates a jump from the supersonic branch of the curve with $A_* = A_t$ to the upper branch of the curve through p_r, A_e.

When the receiver pressure p_r is lowered from $p_r = p_2$, the shock front will move from the throat toward the exit. It will reach the exit for a value $p_r = p_4 > p_6$. In other words, for $p_r < p_4$ no adjustment of the flow to the receiver pressure is possible by a shock in the nozzle. Again a new type of flow pattern must be found to describe what happens under the condition $p_r < p_4$.

Certainly we must now expect that in the nozzle the flow will be the same as that in the ideal case (6). The whole curve $A = A_t f\left(\frac{p}{p_*}\right)$ (precisely, the subsonic branch for $A_t < A < \infty$ and the supersonic branch $A_t < A < A_e$) indicates the flow in the nozzle. Now it is in the jet outside the nozzle that the adjustment to the outside pressure p_r takes place. There will be two types of phenomenon according as $p_4 > p_r > p_6$ or $p_r < p_6$; the intermediate case $p_r = p_6$ is the ideally adjusted continuous flow considered before.

83. Shock patterns in nozzles and jets. To understand the adjustment of the flow in the jet it is better first to revert to the shock patterns occurring within the nozzle for $p_4 < p_r < p_2$ and refine their description. These shock fronts are actually not just simple discs across the nozzle and perpendicular to the nozzle wall. They do not start perpendicularly at the wall; hence the shock front there is oblique, and consequently changes abruptly the direction of the flow, thereby leading to jet detachment. The actual situation is represented in Fig. 7.

At the wall the shock front begins obliquely as a cone and is

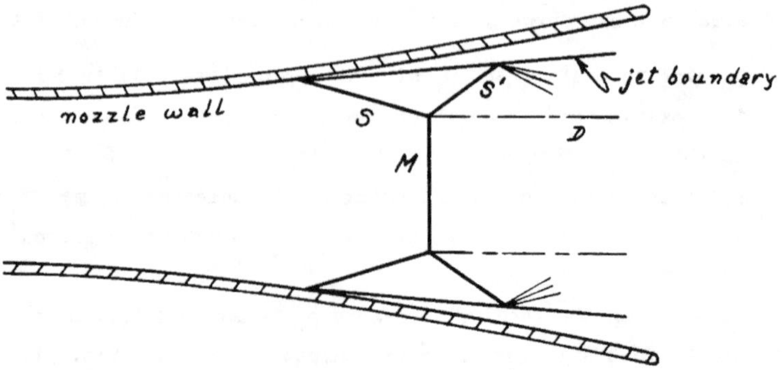

Figure 7
Jet detachment and shock pattern inside a nozzle.

followed by jet detachment. The shock cone is cut off by the "Mach shock disk" perpendicular to the axis (Fig. 7), presenting approximately the picture envisaged in the simplified description of the preceding article. Behind the incident and Mach shock front a conical reflected shock front S' and a discontinuity surface D develops.

When the receiver pressure p_r decreases to the value p_4, the place of detachment moves toward the rim of the nozzle and remains there when p_r becomes less than p_4, while the shock front leaving the rim becomes longer (see Fig. 8). If the receiver pressure has decreased to the value p_6, the strength of the shock leaving the rim becomes zero. On further decrease of the receiver pressure p_r below p_6 a new set of phenomena begins.

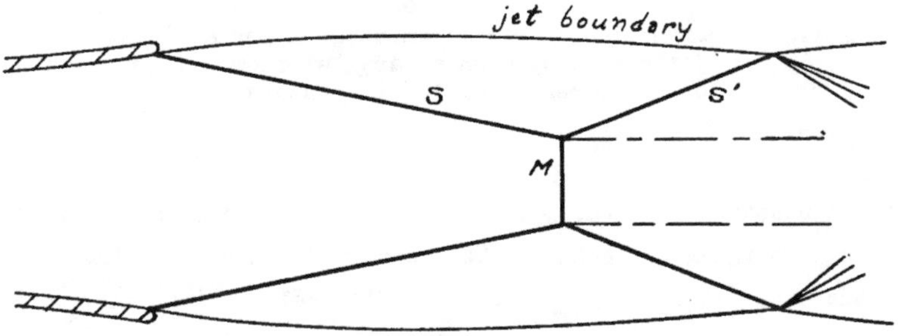

Figure 8
Shock pattern in a jet emerging with pressure less than receiver pressure.

Prandtl's original idea of what happens when the receiver pressure is below the exhaust pressure was the following (see Art. 64, Chapter IV). First suppose that the jet emerges from the nozzle with constant axial velocity. If the flow is two-dimensional the flow pattern consists of two centered rarefaction waves leaving the rim, intersecting each other and being reflected as converging waves

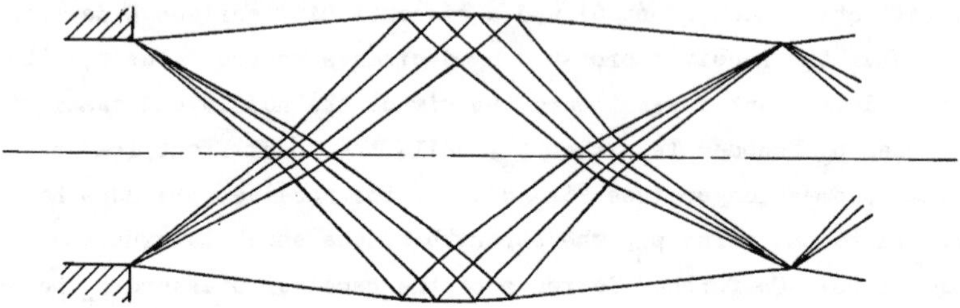

Figure 9
Prandtl's wave pattern assumed for a jet resulting when a parallel flow enters a region of lower pressure.

at the boundaries of the jet (see Fig. 9). Prandtl assumed that these waves converged to a point at the opposite boundary and that from there on the wave pattern repeats itself periodically. A similar pattern was expected for three-dimensional jets and for diverging jets emerging from the nozzle. Experiments, however, contradict this expectation (see Stanton [46]). It is true, naturally, that at the rim expansion waves develop which tend to lower pressure and density to the values in the receiver, but somewhere on the outer border of the rarefaction waves a shock front develops which cuts across the rarefaction wave and "intercepts" or "stops" it. The flow patterns that may result are indicated in Figs. 10a and 10b. These intercepting shock fronts appear to be a natural continuation of the shock fronts which begin at the rim if $p_r > p_6$. Even in the ideal case where the receiver pressure equals the exhaust pressure, $p_r = p_6$, intercepting shock fronts develop in the jet.

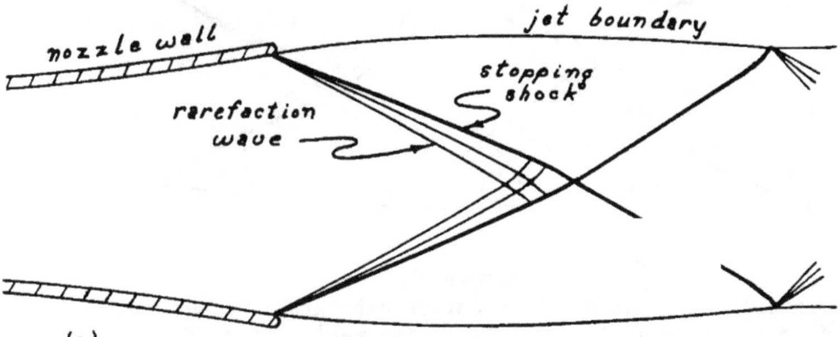

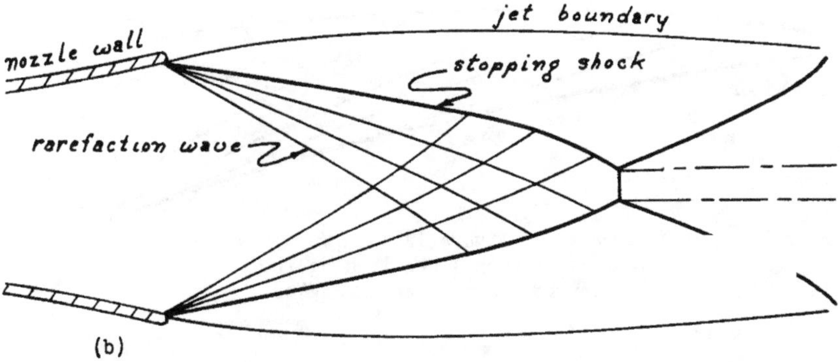

Figures 10a and 10b
Shock patterns in a jet emerging with pressure greater than receiver pressure.

That such intercepting or stopping shocks must occur can be made plausible. First of all it seems that even for a two-dimensional jet emerging with constant velocity, Prandtl's pattern as shown in Fig. 9 is not a correct description; there is no reason why the contracting wave resulting from the reflection of the incident wave at the jet boundary should be a mirror image of the incident wave and converge to a point at the opposite boundary. It seems (although this is not obvious) that the converging reflected wave contracts faster than if it were a mirror image of the incident wave and that consequently there results a shock front in the form of a truncated

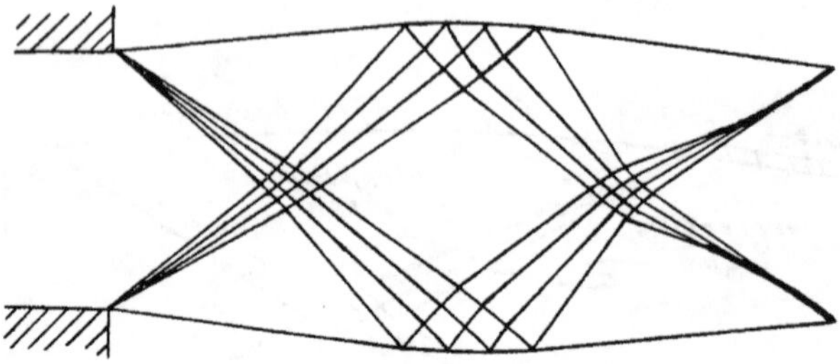

Figure 11
Probable wave pattern in a jet resulting when a parallel flow enters a region of lower pressure (contrast Fig. 9).

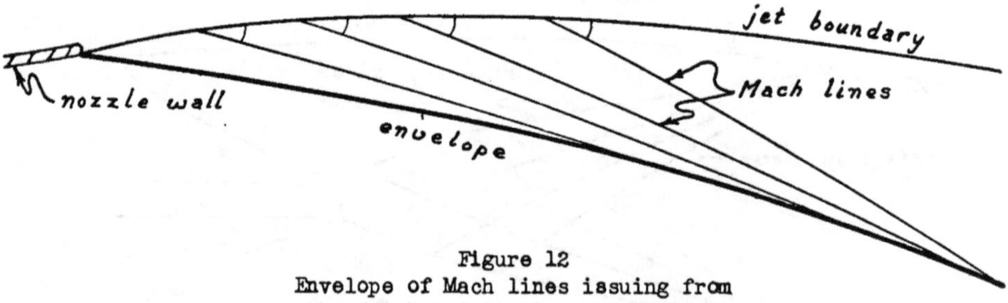

Figure 12
Envelope of Mach lines issuing from the jet boundary at equal angles.

cone (Fig. 11). When the ratio of the pressure of the emerging jet to the outside pressure is increased, this shock front will eventually form a full cone and, on further increase, configurations similar to those shown in Fig. 10a or 10b will ensue, depending on the Mach number of the emerging jet (see Fig. 3 in Hartmann and Lazarus [52]).

This situation is even more pronounced in a jet emerging from a <u>nozzle</u> mouth when the receiver pressure is less than p_6. The flow continues uninfluenced by the state in the receiver until the first Mach line issuing from the rim, i.e., the inner border of the rarefaction wave is met. Owing to the divergence of the jet, the pressure decreases along the axis. The decrease of pressure is considerable when the Mach number $\frac{q}{c}$ of the emerging jet is noticeably greater than unity.* Across the rarefaction wave the pressure further decreases to atmospheric pressure at the rim, hence to below atmospheric pressure farther out. In other words, the pressure below one atmosphere at the outer border of the rarefaction wave while it equals one atmosphere at the jet boundary. Consequently there is a pressure gradient acting from the jet boundary toward the interior of the jet. Clearly this pressure gradient will curve the jet inward. All the Mach lines issuing from the boundary make the same angle with the boundary, since at the boundary the pressure, hence also c and q, remains constant. These Mach lines thus tend to converge and very likely would have an envelope if they were not intercepted by a shock front. To prevent the envelope singularity, a "stopping" shock is therefore necessary (see Fig. 12).

Various shock patterns may result, depending on the degree of divergence of the exhaust flow, its Mach number, M_e, and the ratio of its pressure p_e to the outside pressure p_a. A most typical case is that shown in Fig. 8 for $p_r > p_6$. The pattern is similar when $p_r < p_6$ except that the shock front does not begin at the rim but emerges as a stopping shock. It involves conical "incident" and "reflected" shock fronts connected by a Mach disk shock front perpendicular to the axis. The characterizations "incident" and "reflected" are used only to identify these parts of the shock pattern with those occurring in obliquely impinging shock waves. In a certain sense one may say that in the present problem the "reflected" shock is the primary phenomenon, the "incident" one being determined by it. (In certain cases of jets with weak divergence only a section of the "reflected" shock wave has been observed, Mach disk and "incident" shock being absent (see Fig. 11).) It should

---

* This follows from the relation

$$\frac{dp}{p} = \gamma\left(\frac{q}{c}\right)^2 \frac{dq}{q} .$$

also be noted that the "reflected" shock is of the strong variety, an occurrence never observed in the reflection of impinging shock waves.

The configuration shown in Fig. 8 is observed, for instance, for $M_e \sim 3$, if the half-angle of divergence is greater than 15°. The pressure in front of the Mach disk is found to be of the order of magnitude $0.1\, p_a$ to $0.03\, p_a$, hence very low. The strength of the Mach shock is very great, its excess pressure ratio being of the order of magnitude 20 to 50.

A few words about the continuation of the jet may be added. The jet boundary will curve inward up to the place where the stopping shock meets the boundary. This shock front will there be reflected as a rarefaction wave and the jet boundary will diverge again. The whole process repeats itself. Due to the action of viscosity at the jet boundary this periodic jet pattern will eventually be blurred.

84. Thrust. Exhaust flow out of a nozzle is an essential element of a rocket motor. The burnt gases which are formed in the combustion chamber under high pressure acquire a considerable momentum when they are ejected through the nozzle. Accordingly, as a reaction to this momentum flux, a thrust results which acts against the rocket in the direction opposite to that of the exhaust flow. The total thrust against the rocket can easily be expressed in terms of the quantities characterizing the exhaust flow, and conditions can be formulated for the shape of the nozzle such that a maximum thrust is provided.

It is customary to define the total thrust F as the difference

$$F = F_i - F_a \tag{13}$$

of the internal thrust F_i, resulting from the pressure acting against the wall of the chamber and the nozzle, and the external counterthrust F_a that would result if atmospheric pressure were acting against the outer surface of the body in which the nozzle is imbedded.

To evaluate the thrust we consider the surface S through the exit rim on which the speed and hence the pressure is constant. The internal thrust F_i (counted positive when acting against the stream) is equal to the sum of the axial component of the momentum M transported through S to the outside per unit time and of the resultant pressure force, F, exerted against the surface S from the inside.*

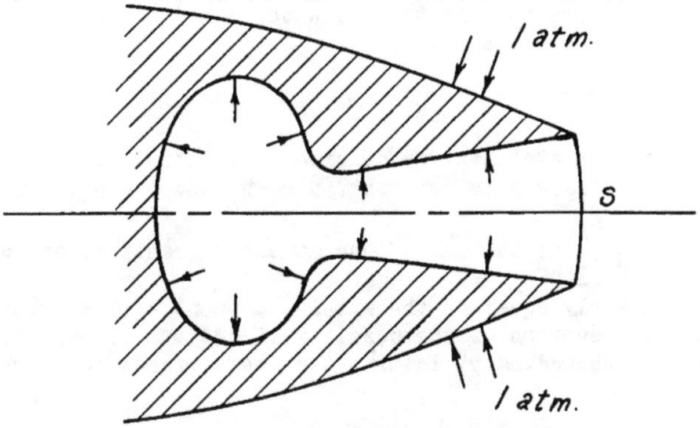

Figure 13
Forces contributing to the thrust

We denote by A the projection of the surface S on a plane perpendicular to the axis; then $P = pA$ and the internal thrust is

(14) $$F_i = M + pA.$$

The external thrust is clearly

(15) $$F_a = p_a A$$

counted positive when acting in direction of the stream.

We consider the total thrust

(16) $$F = F_i - F_a = M + (p - p_a)A$$

a function of the position of the rim, imagining that the nozzle may be continued or cut off at various places. In particular, we

---

* I.e., the total pressure force, $F_i - P$, exerted against the volume of gas enclosed by chamber, nozzle and surface S, equals the momentum M transported in unit time through S.

are interested in discovering for which position of the rim the total thrust is a maximum. The answer can be given completely.

The total thrust is a maximum when the nozzle is cut off at such a place that the pressure at the exit rim just agrees with the outside pressure. In that case the total thrust is just given by the momentum transport

(17) $$F_{max} = M.$$

To prove this statement we consider two different surfaces S on which the pressure p is constant, $p = p_1$ and $p = p_2$. The change $\left[M + p A\right]_{p_1}^{p_2}$ of the sum of the momentum transport and pressure force is clearly equal to the axial component of the pressure force against the section of the nozzle wall cut out between the two surfaces S. Accordingly, letting the two surfaces coalesce, we find

$$d(M + pA) = pdA.$$

Consequently

$$dF = (p - p_a)dA.$$

It is thus shown that F is an extremum when $p = p_a$, or $dA = 0$. It can be shown that at the throat, where $dA = 0$, F is a minimum, while F is a maximum when $p = p_a$.

The maximum thrust so determined depends on the shape of the nozzle contour. This shape can be so determined that the maximum thrust is a maximum for all nozzles delivering the same mass flux.

The position of the exit rim of a fixed nozzle contour for which the maximum thrust is obtained may be characterized by a subscript ($_m$). The surface S through this rim is S_m; the speed q_m on S_m is solely determined by formulas (4), (6), and (7) through the condition that the pressure p_m on S_m equals p_a. The maximum thrust is given by

(18) $$F_m = M_m = q_m \int_{S_m} \cos\theta \, dG$$

where θ is the angle of the flow direction with the axis and dG the element of the mass flux per unit time G. Clearly $F_m \leq q_m G$ and $F_m = q_m G$ only if $\theta = 0$ on S_m. In other words, the maximum thrust is a maximum for fixed mass flux G if the exhaust velocity is constant and has axial direction.

85. Perfect Nozzles. A nozzle which produces a constant axial exhaust flow may be called a perfect nozzle. Such perfect nozzles can be designed without difficulty. As a matter of fact, whenever a diverging exhaust flow is given, it is possible, by re-routing only a section of it, to make the flow "perfect", i.e., to guide it so that it eventually acquires constant axial velocity. Every streamline of such a perfect flow yields a perfect nozzle. The possibility of constructing a perfect flow has already been indicated by Prandtl and Busemann [3],[45],[47] (see also [49]).

Perfect nozzles can be constructed so as to produce any desired exhaust velocity; that is, expressed in dimensionless terms, the Mach number of the exhaust flow, or what is equivalent, the ratio of chamber to exhaust pressure can be prescribed. The first step in the construction consists in securing an exhaust flow, the basic "flow" F_o, which leads at least to the desired exhaust velocity. The re-routing process for a two-dimensional exhaust flow is so simple that it may be described briefly. First the point A_o on the axis should be found where the basic

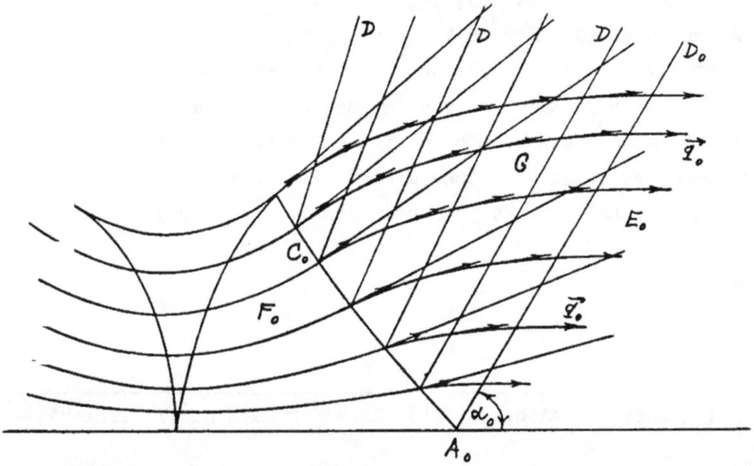

Figure 14
Construction of perfect two-dimensional nozzle.

exhaust flow F_0 attains the desired exhaust velocity q_0. Through the point A_0 two lines are drawn, the backward Mach line C_0 which is determined by the flow F_0, and the straight line D_0 which would be the forward Mach line of a flow E_0 with constant parallel velocity q_0 (pressures p_0 and sound speed c_0 being determined through the condition of isentropic change). The angle between the line D_0 and the axis is just the Mach angle α_0 of the flow E_0. Up to the line C_0 the original flow F_0 will be retained; in the sector G between C_0 and D_0 the flow will be changed, and beyond D_0 it will be parallel with the constant axial velocity q_0. To determine the new flow in the sector G, the known directions of the Mach lines of the flow F_0 should be marked on the line C_0. Straight lines, D, should then be drawn from C_0 in these directions so as to cover the sector G. Further, the known direction of the flow F_0 on the line C_0 should be marked and each such direction is to be transplanted parallel to itself along the lines D. Thus the directions of the new flow F in the sector G are determined. Through integration of this field of directions, beginning at C_0, the streamlines of the new flow F are obtained. Beyond the line D_0 the flow is to be continued with constant axial velocity. We note that the streamlines so constructed suffer a change of curvature on crossing the Mach lines C_0 and D_0.

Perfect nozzles can also be designed for three-dimensional flow with axial symmetry. Although the construction is no longer so simple as for two-dimensional flow, it has been possible to carry it out.*

---

* See Frankl (quoted in Kisonko [48]), Busemann [47] and Friedrichs [49].

B. Conical Flow

86. Qualitative Description. The second type of problem treated in this chapter, that of "conical flow", permits a rather far-reaching analysis on the basis of the differential equations. It concerns steady, isentropic, irrotational flow with symmetry about the x-axis and under a further assumption, that the flow is conical, i.e., that the quantities u, ρ, p, q retain constant values on cones (considered infinite) with a common vertex, the origin. Flow satisfying this condition may occur, for instance, at the conical tip of a projectile opposed to a supersonic stream of air.

The flow against a cone is analagous to the flow against a wedge and, as in the case of a wedge, two cases must be distinguished. If the cone angle is not too large, the deflection of the flow is achieved by a shock front which begins at the tip of the cone and has the shape of a straight cone (Fig. 14a). If, however, the cone angle exceeds a certain extreme value (Fig. 14b), no such conical shock front is possible. Instead, a curved shock front stands ahead of the cone. Only the first

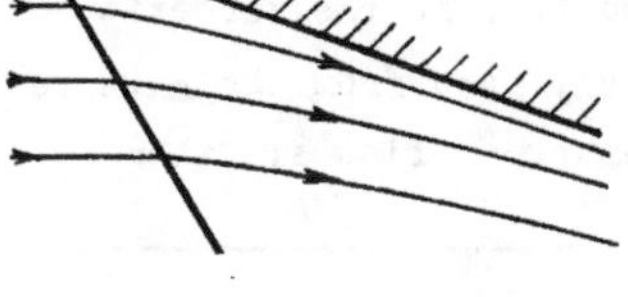

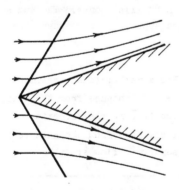

Figure 14a
Conical shock front and conical flow resulting from supersonic flow against a cone with a sufficiently small angle.

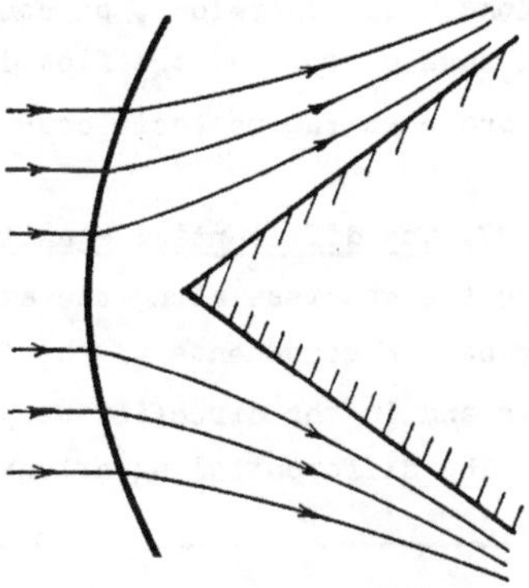

Figure 14b
Curved shock front in supersonic flow against a cone with a large angle.

case can be handled on the basis of the assumption that the flow is conical. We therefore confine ourselves to this case.

In reality projectiles are not represented by infinite cones; they have a conical tip and then taper off, e.g., into a cylindrical shape of finite length. Thus, farther back, the wave from the conical tip interacts with other waves, such as expansion waves coming from the bend of the projectile.* It is worth while noting that, in the case of a shock wave standing ahead of the projectile, the distance, under otherwise equal conditions, is the greater the farther the cone extends before tapering off.

Returning now to the idealized case of a strictly conical flow, we may describe the situation qualitatively as follows. Ahead of the shock front the air is in a constant state flowing in the direction of the axis with constant velocity. Since the shock front is a straight cone making everywhere the same angle with the incident flow, the state behind it is also constant and it is therefore clear that the flow is isentropic behind the shock front. Moreover, it can be continued so as to satisfy the basic assumption that the flow is conical. The state of the air beyond the shock cone will, therefore, be constant on co-axial cones. The angle between such a cone and the flow direction approaches zero when this cone approaches the obstacle cone.

87. <u>The differential equations</u>. For a mathematical treatment,** let x be the abscissa along the axis, r be the distance from the axis, u and v be the components of the flow velocity $\vec{q}$ in the direction of the axis and in the direction perpendicularly away from the axis respectively. The differential equations for isentropic flow are then

---

* For "two-dimensional projectiles" a rather complete theory is possible on the basis of the material in Chapter IV. See also the paper by Epstein [35].

**See Busemann [59] and Taylor and Maccoll [57], [58].

(19) $$v_x = u_r$$

(20) $$r(\rho u)_x + (r\rho v)_r = 0$$

where ρ is given by the relation

(21) $$\frac{\rho}{\rho_*} = \left(\frac{c}{c_*}\right)^{\frac{2}{\gamma - 1}}$$

and Bernoulli's law

(22) $$\left(\frac{c}{c_*}\right)^2 = \frac{1 - \mu^2\left(\frac{q}{q_*}\right)^2}{1 - \mu^2}, \quad q^2 = u^2 + v^2$$

Inserting (21) and (22) into (20) one has

(23) $$\left(1 - \frac{u^2}{c^2}\right)u_x + \left(1 - \frac{v^2}{c^2}\right)v_r + \frac{v}{r} - 2\frac{uv}{c^2}v_x = 0.$$

The basic assumption of conical flow now implies that u, v and hence c depend only on the ratio

(24) $$t = \frac{x}{r}.$$

Equation (19) then becomes

(25) $$v_t + tu_t = 0$$

while (23) is reduced to

(26) $$\left(1 - \frac{u^2}{c^2}\right)u_t - \left(1 - \frac{v^2}{c^2}\right)tv_t + v - 2\frac{uv}{c^2}v_t = 0.$$

Equations (25) and (26) are a pair of differential equations of the first order for the two functions u and v of t. Clearly this pair is equivalent to one equation of the second order for one function.

This equation of second order assumes a form which is particularly amenable to treatment when v is introduced as function of u. From (25) we have

$$t = -v_u \tag{27}$$

Differentiation of this relation with respect to t yields

$$u_t = -\frac{1}{v_{uu}} \tag{28}$$

This relation together with (27) and (25) gives

$$v_t = -\frac{v_u}{v_{uu}} \tag{29}$$

Insertion of equations (27), (28) and (29) into (26) gives

$$\left(1 - \frac{u^2}{c^2}\right) + \left(1 - \frac{v^2}{c^2}\right)v_u^2 - vv_{uu} - 2\frac{uv}{c^2}v_u = 0$$

or

$$vv_{uu} = 1 + v_u^2 - \frac{(u + vv_u)^2}{c^2} \tag{30}$$

Every section of a solution of equation* (30) gives a flow provided that

---

* It may be mentioned that Busemann gives an elegant geometric interpretation of equation (30):

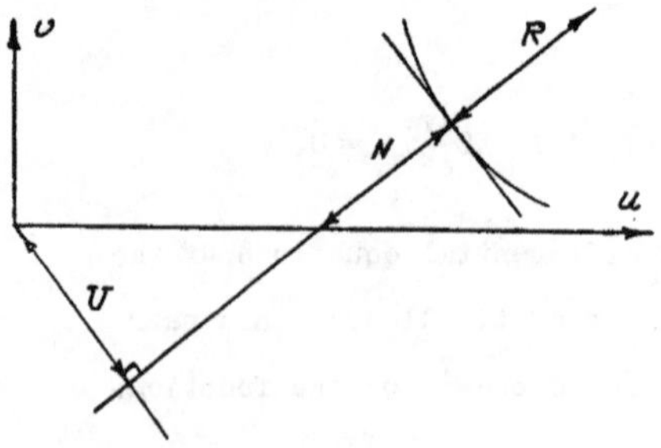

Figure 16

$$R = \frac{N}{1 - \frac{U^2}{c^2}} \tag{31}$$

R being the radius of curvature and the meaning of N and U being obvious from Fig. 16.

condition

$$(32) \qquad v_{uu} \neq 0$$

is satisfied, because then x and r can be introduced as independent variables by $v_u = -t = -x/r$. Thus the ray to which values of u and v are to be attached is determined. The direction of this ray in the x,y-plane is evidently normal to the curve $v = \dot{v}(u)$ at the point (u,v) in the hodograph plane.

The flows so obtained are in a certain way analagous to centered simple waves for two-dimensional flows. However, while in the case of two-dimensional steady flow the simple waves are represented in the hodograph plane by two families of fixed characteristics (epicycloids), the images of the special flows considered here in the hodograph plane correspond to a greater variety of curves, namely, a whole family through each point.

88. <u>Conical shocks.</u> The relations governing the transition through a <u>conical shock</u> are the same as for the plane oblique shock; the curvature of the shock cone does not enter. When the shock cone is a straight cone, as is assumed, the jumps of u,v,p, and of the entropy are constant along each ray when the assumption of conical flow is satisfied on one side; consequently this assumption remains satisfied on the other side. The flow may continue as a conical flow with constant entropy after crossing the shock. In other words, the assumption of proper conical shocks is compatible with the basic assumption.

Suppose a flow characterized by p_0, ρ_0, u_0, v_0 crosses such a a conical shock. (It is to be noted that this can occur only if the speed $q_0 = \sqrt{v_0^2 + v_0^2}$ is supersonic, i.e., if $q_0 > c$). The velocity $q_1 = (u_1, v_1)$ immediately past the shock front is located on the loop of the strophoid in the u,v-plane. The inclination of the ray which generates the shock cone is perpendicular to the straight connection between (u_0, v_0) and (u_1, v_1). The positions of the cones corresponding to the cases (a): $v_1 > v_0$ and (b): $v_1 < v_0$ are indicated in Fig. 17.

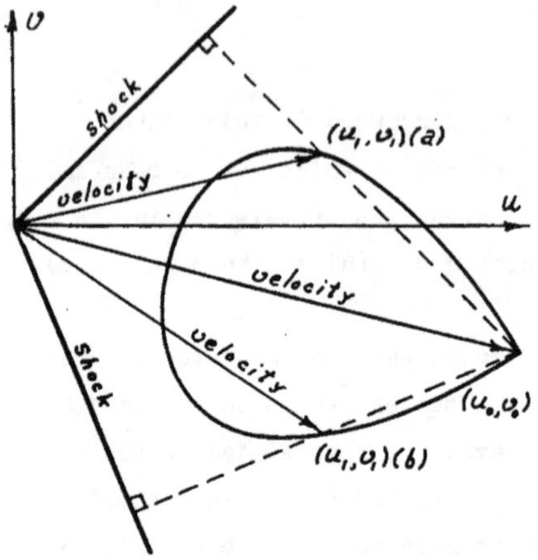

Figure 17
Indicating transition through conical shock front for cases (a): $v_1 > v_o$ and (b): $v_1 < v_o$.

When the flow on either the front or the back side of the shock is to be continued according to differential equation (30), the slope of the u,v-curve is to be so determined that the ray given by (27) coincides with the shock. Since this ray is to be normal to the u,v-curve on the one hand and perpendicular to the straight segment connecting (u_0, v_0) with (u_1, v_1) on the other hand, the u,v-curve should begin or enter in the direction of this segment. The slope of the u,v-curve is thus given by

$$v_u = \frac{v_1 - v_0}{u_1 - u_o} . \tag{33}$$

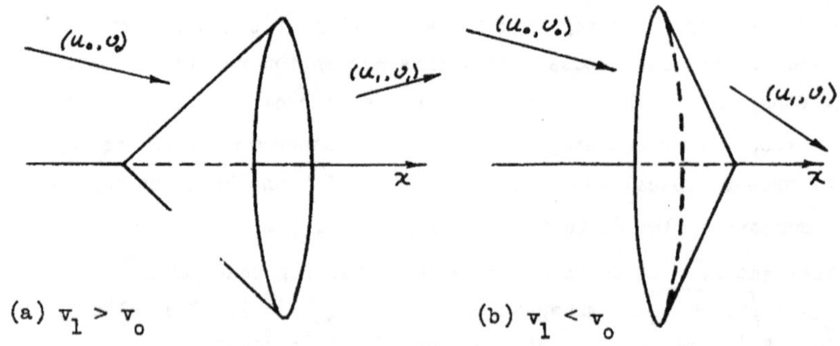

Figure 18
Flow through conical shock fronts.

The discussion of conical shock fronts by Busemann and by Taylor and Maccoll is restricted to case (a) with $u_0 = q_0 > 0$ and $v_0 = 0$. This case (Fig. 18a) occurs when a constant axial flow is deflected by a conical projectile. We shall indicate briefly how Busemann treats this problem. Through the shock transition relations the flow velocity (u_1, v_1) past the shock is given (observe that the third transition relation guarantees that the Bernoulli constant $\frac{1}{2}\hat{q}^2$ is the same before and after the shock). A solution of equation (30) is to be found whose graph passes through the point (u_1, v_1). The slope v_u of this curve is given by (33). The solution is now to be so continued that $t = x/r$ increases, i.e., in view of (27) v_u decreases up to a point at which the flow and the ray have the same direction, i.e., where $v/u = x/r$, or where the normal passes through the origin; such a point may be called an end point. This end point depends on the choice of the point (u_1, v_1) on the strophoid. The manifold of endpoints that can be reached from $(q_0, 0)$ forms a curve which Busemann calls the "apple curve" in view of its peculiar shape, see Fig. 19.

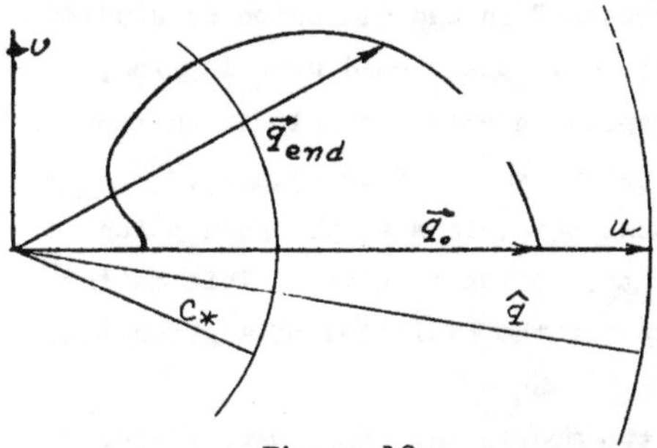

Figure 19
Apple curve showing all possible end velocities that can be reached after crossing a conical shock front from a state with a given incoming velocity.

In this procedure the shock is prescribed and the end direction is found. If the end direction is prescribed one may find the corresponding point on the apple curve by intersecting it with the appropriate ray through the origin. In general, there will be two intersections of which the one corresponding to the weaker shock is likely to occur in reality.*

---

* It may be mentioned that in the procedure of Taylor and Maccoll one begins with the end direction, the shock then being found by following the solution of (26) backwards. This procedure has advantages when single cases are to be investigated.

The values of pressures and angles calculated on the basis of the preceding considerations agree exceedingly well with experimental values (see Taylor and Maccoll [57], [58]).

C. Spherical Waves

89. General remarks. Spherical wave motion* is obviously a subject of basic importance for the study of explosion waves in water, air and other media. In spherical motion the velocity is radial and its magnitude as well as that of density, pressure, temperature, and entropy depends only on the distance r from the origin and on the time t. Such motion might be considered in a certain sense as somewhat analagous to one-dimensional motion in a tube under the influence of a piston. In the three-dimensional space the piston is replaced by an expanding (or contracting) sphere which impresses a motion on the medium inside or outside.

The simplest model would be that of a "spherical piston" pushing at constant velocity into an infinite surrounding medium. Such a model corresponds to the uniform "piston motion" in one dimension as studied in Chapter III, in particular in Art. 41. One should bear in mind, however, that in three-dimensional space an energy supply at an increasing rate is required to maintain constant speed of the piston.

In better agreement with actual situations is the assumption that the total energy available for the motion is given. This is the case for spherical blast waves caused by the explosion of a given mass of explosive.

While in the first of these two models the shock wave racing ahead of the piston has constant speed so that the shock conditions are compatible with the assumption of isentropic flow on both sides of the

* With slight modifications the following considerations apply also to cylindrical waves.

discontinuity, this is no longer true of blast waves. In the latter the strength of the shock, and hence the change of entropy, rapidly decreases so that behind the shock front the flow is no longer isentropic. Moreover, in blast waves the air or water, after having crossed the shock front and having thereby undergone compression, will rapidly expand again to a pressure in general even below that in front of the shock wave. This suction phase is an important feature of motion caused by explosions.

A phenomenon of major importance is that of reflection of spherical shock fronts; a contracting spherical wave preceded by a shock front may be "reflected" at the center with the result of enormous pressure increase behind the reflected shock front.*

At the present state of knowledge all that can be done along the lines of mathematical analysis is to find and to discuss some particular solutions of the differential equations of spherical waves which are approximately in agreement with the additional conditions of the problems. One may hope that these solutions display at least qualitatively important features of reality. It is remarkable that such an unambitious approach seems to be sufficient to lead to a certain degree of understanding and control of actual phenomena.

90. Analytical formulations. Assuming that the velocity is radially directed and that the radial component of velocity u, the pressure p and the density depend only on the distance r from the center at the time t, the differential equations are (see II (F), Art. 8).**

(34) $$u_t + uu_r + \frac{1}{\rho}p_r = 0,$$

(35) $$\rho_t + u\rho_r + (u_r + \frac{2u}{r}) = 0,$$

(36) $$(p\rho^{-\gamma})_t + u(p\rho^{-\gamma})_r = 0,$$

*Such shock reflection in three dimensions with half-spherical, cylindrical or conical symmetry probably plays a decisive role in hollow charge effects.

**We have used here the notation r for the distance from the origin instead of the x employed previously.

assuming that the medium is polytropic with the adiabatic exponent γ. The third equation expresses the fact that the entropy is constant along the path of a particle. It is not assumed that the entropy is constant throughout since, as stated before, the entropy does not in general remain constant behind a shock front. If the head of the wave is given as a function

(37) $$r = R(t),$$

the total energy carried by the wave motion is expressed as

(38) $$E = 4\pi\int_0^R \left\{\frac{1}{2}\rho u^2 + \frac{1}{\gamma - 1}p\right\} r^2 dr.$$

E is clearly a function of the time t.

Another important quantity, the impulse I per unit area received by a section of the surface of the sphere at distance r, is given by

(39) $$I = \int_T^\infty p\,dt$$

where $T = T(r)$ is the time at which the wave front arrives at the place r. $T(r)$ is connected with $R(t)$ through $r = R(T(r))$. Clearly, I is a function of r.

91. Special solutions. According to classical procedure, one may obtain particular solutions of the differential equations by assuming a specific form for the solution to reduce the problem to one involving ordinary differential equations. Thus solutions are obtained which have been called progressing waves.* These are solutions, conveniently assumed in the form

(40) $$u = t^\beta \xi U(\xi),\ \rho = t^\delta P(\xi),\ \frac{p}{\rho} = t^\varepsilon \xi^2 T(\xi)$$

where ξ is the combination

(41) $$\xi = rt^{-\alpha}.$$

---

* Concerning a more general concept of "progressing wave" see, for instance, Courant-Hilbert [12], Vol. II, p. 448.

In other words, a progressing motion is a special motion for which the quantities $ut^{-\beta}$, $\rho t^{-\delta}$, $pt^{-\varepsilon-\delta}$ appear constant for an observer who moves on a path given by $rt^{-\alpha}$ = constant. The exponents α, β, δ, ε should be so adjusted that upon insertion of (40) and (41) in (34), (35), (36), equations result which involve only the variable ξ and no longer the variables r and t explicitly. One immediately verifies that to this end one must set

(42) $$\beta = \alpha - 1, \quad \varepsilon = 2\beta$$

The equations for U, P, T are then

(34') $$(U-\alpha)(\xi U' + U) + \beta U + \xi T' + 2T + \frac{\xi P'}{P} T = 0$$

(35') $$(U-\alpha)\xi P' + \delta P + (\xi U' + 3U)P = 0$$

(36') $$(U-\alpha)\xi T' - (\gamma-1)(U-\alpha) + \frac{\xi P'}{P} T + (2\beta - (\gamma-1)\delta)T = 0.$$

It is interesting that after elimination of $\frac{\xi P'}{P}$ by (35'), equations (34') and (36') can be reduced to one equation of first order for T as a function of U.*

The equations (34'), (35'), (36') are, of course, amenable to a numerical solution.

When the head of the wave is given by

(37') $$r = R(t) = \Xi t, \text{ or } \xi = \Xi,$$

Ξ being a constant, we obtain for the energy

(38') $$E = 4\pi t^{5\alpha+\delta-2} \int_0^{\Xi} \left\{\frac{1}{2}U^2 + \frac{1}{\gamma-1}T\right\} P \xi^4 d\xi$$

and for the impulse per unit area

(39') $$I = \frac{1}{\alpha} r^{\frac{\delta-1}{\alpha}+2} \int_0^{\Xi} \frac{P(\xi)T(\xi)}{\xi^{\frac{\delta-1}{\alpha}+1}} d\xi .$$

* See Guderley [64].

92. Discussion of Special Cases. We shall discuss the simplifications resulting from several special assumptions.

(a) If the flow is isentropic, implying constant strength of the shock ahead of it, then relations (40) and (42) require that

$$\delta = \frac{2}{\gamma - 1}\beta \ . \tag{43}$$

If in particular $\delta = \beta = 0, \alpha = 1$, the head of the wave $r = \Xi t$ moves with constant velocity Ξ and ρ and p are constant behind it. Such wave motion is therefore compatible with a constant shock front.

As mentioned before, a wave of this type will result if a sphere is suddenly expanded with constant velocity. After crossing the shock front, every air particle acquires the same pressure, density, entropy and velocity. Thereafter, as can be shown, the air particles are further compressed and accelerated and their velocity approaches asymptotically that of the expanding sphere.

(b) Except for the case just mentioned, a shock at the head of the wave is not exactly compatible with a progressing wave. Strong shocks, however, are compatible to a good approximation, if the exponents β and δ are properly related.

Denoting by ρ_0 the density ahead of the wave, and setting the pressure ahead of the wave equal to zero,* the shock transition conditions reduce to

$$\rho = \mu^{-2}\rho_0, \quad p = (1 - \mu^2)\dot{R}^2\rho_0, \quad u = (1 - \mu^2)\dot{R}, \tag{44}$$

as can be inferred by setting $p_0 = 0, \rho_0 = 0$, $u_0 = 0$, and $\xi = \dot{R}$ in

---

* This is the simplifying approximation corresponding to the assumption of a strong shock.

IV (i'), (ii_N'), (iii_N'). Insertion shows that these relations are compatible with (40), (42), and (37') only if

$$(45) \qquad \delta = 0,$$

i.e., if the density remains constant on the paths $r = \xi t^{\alpha}$. For the wave motion behind the shock front one then obtains the boundary values

$$(46) \quad P(\Xi) = \mu^{-2}\rho_0, \; T(\Xi) = \mu^2(1-\mu^2)(\beta+1)^2,$$

$$U(\Xi) = (1-\mu^2)(\beta+1).$$

A situation of particular interest arises if the shock wave contracts toward the origin and is eventually reflected by another progressing wave preceded by a strong shock. Such an occurrence can be expressed in terms of progressing waves of the type considered here only if $\alpha = .717$ (or $\alpha = .834$ for cylindrical waves).* It is very significant (see [64], Fig. 4) that the pressure past the reflected shock front is about 26 times the pressure behind the incident shock front (for air, $\gamma = 1.4$) as compared with a 17-fold increase for cylindrical motion and an 8-fold increase for one-dimensional motion.

(c) The condition that the energy remains constant leads by (38') to the condition

$$(47) \qquad \delta = -5\alpha + 2.$$

If, in addition, the wave is to possess a strong shock at its head, so that $\delta = 0$, we have

$$(48) \qquad \beta = -\frac{3}{5}, \quad \alpha = \frac{2}{5}, \quad \varepsilon = -\frac{6}{5}.$$

---

* See Guderley [64].

The motion of the shock front is then given by

$$r = \Xi t^{\frac{2}{5}} . \tag{49}$$

The pressure behind the shock,

$$p = \frac{4}{25}(1-\mu^2)\,\rho_o \Xi^2 t^{-\frac{6}{5}} = \frac{4}{25}(1-\mu^2)\,\rho_o \Xi^5 R^{-3} , \tag{50}$$

approaches zero as $t \to \infty$. Consequently the assumption that the shock is strong will eventually be violated. As long as this assumption is valid, however, the solution represents a progressing blast wave. G. I. Taylor, who first recognized its existence, has carried out the solution numerically and has been able to draw important conclusions from the results (see [63]), although actual blast waves are in general not of this simple, "progressing" type.

The difficulties of determining non-progressing spherical waves are very great and for that reason inferences from various approximate treatments have been attempted. The "incompressible approximation" arises when one lets γ, and accordingly c, become infinite, while ρ remains constant. For water, with $\gamma = 7$, this appears to be acceptable. The "sonic approximation" results when the deviation from the state at rest is small so that only linear terms in these deviations need be considered.*

Finally it may be mentioned that certain conclusions can be drawn from the differential equations (1), (2), (3) by a purely dimensional analysis. Any solution, $\tilde{u}(r,t)$, $\tilde{\rho}(r,t)$, $\tilde{p}(r,t)$, leads to a variety of other solutions

---

* For such approximate treatments see Bethe, Kirkwood [65].

$$(51) \qquad u = c_0\tilde{u}\left(\frac{r}{r_0}, \frac{t}{t_0}\right), \quad \rho = \rho_0\tilde{\rho}\left(\frac{r}{r_0}, \frac{t}{t_0}\right), \quad p = p_0\tilde{p}\left(\frac{r}{r_0}, \frac{t}{t_0}\right)$$

when r_0, t_0, c_0, ρ_0, p_0 are any fixed quantities (of obvious dimensions) satisfying

$$(52) \qquad r_0 = c_0 t_0, \quad p_0 = \rho_0 c_0^2 .$$

One may choose p_0 as the pressure, c_0 as the sound speed ahead of the shock wave. In the case of a blast wave with the energy E_0 one may set

$$(53) \qquad r_0 = \left(\frac{E_0}{p_0}\right)^{1/3} .$$

Then one finds for the impulse per unit area (see (39), Art. 91)

$$(54) \qquad I = p_0 t_0 \tilde{I}\left(\frac{r}{r_0}\right) = \frac{\left(E_0 p_0^2\right)^{1/3}}{c_0} \tilde{I}\left(\frac{r}{r_0}\right).$$

In conclusion, it might be emphasized again that the theory of flow in three dimensions is still in a state where one can proceed only by groping for such clues as may come from typical examples which can be handled by some special device. There is much need for, and some prospect of, progress in this field.

BIBLIOGRAPHY

For a comprehensive bibliography of literature concerning flow of compressible fluids, see:

[1] Michel, L. R., Bibliography on flow of compressible fluids. Dept. of Mechanical Engineering, Massachusetts Institute of Technology, 1942-43.

General references of particular interest and importance are:

[2] Taylor, G. I., and Maccoll, J. W., "The mechanics of compressible fluids." Vol. III, W. F. Durand's Aerodynamic Theory, Division H, Julius Springer, Berlin, 1935.

[3] Busemann, A., "Gasdynamik". Handbuch der Experimentalphysik, Vol. 4, Akademische Verlagsgesellschaft, Leipzig, 1931.

[4] Ackeret, J., "Gasdynamik". Handbuch der Physik, Vol. 7, Springer, Berlin, 1927.

Some surveys of the field in the form of lectures are:

[5] Taylor, G. I., "Recent work on the flow of compressible fluids." Journal of London Mathematical Society, Vol. 5, 1930, pp. 224 - 240.

[6] Prandtl, L., "Allgemeine Überlegungen über die Strömung zusammendrückbarer Flüssigkeiten." Zeitschrift für angewandte Mathematik und Mechanik, Vol. 16, 1936, pp. 129 - 142, and literature given there.

[7] von Kármán, Th., "The engineer grapples with non-linear problems." Bulletin of the American Mathematical Society, Vol. 46, No. 8, 1940, pp. 615 - 683, and literature given there.

Short introductions into the field of compressible fluid flow may be found in:

[8] Lamb, H., Hydrodynamics. Cambridge, 1932.

[9] Milne-Thompson, L. M., Theoretical Hydrodynamics. Macmillan, 1938.

[10] Friedrichs, K. O., Lecture notes, "Fluid dynamics and applications." New York University, 1941.

[11] von Mises, R., and Friedrichs, K. O., Lecture notes, "Fluid Dynamics". Brown University, 1941.

For the theory of characteristics of differential equations developed and employed in Chapter II, see:

[12] Courant, R., and Hilbert, D., Methoden der mathematischen Physik, Vol. II. Interscience Publishers, Inc., New York, 1943.

For the basic laws of fluid motion with consideration of thermodynamics, see:

[13] Goldstein, S., Modern Developments in Fluid Dynamics. Aeronautical Research Committee, Fluid Motion Panel, Clarendon Press, Oxford, 1938. Two Volumes.

For the theory of non-linear waves in one-dimensional motion developed in Chapter III, see:

[14] Riemann, B., "Über die Fortpflanzung ebener Luftwellen von endlicher Schwingungsweite." (On the propagation of plane air waves of finite amplitude.) Abhandlungen d. Gesellschaft der Wissenschaften zu Göttingen, Math.-physikal. Klasse, Vol. 8, 1860, p. 43, or Gesammelte Werke, 1876, p. 144.

[15] Hugoniot, H., "Sur la propagation du mouvement dans les corps et specialement dans les gaz parfaits." (On the propagation of disturbance through substances and especially through perfect gases.) Journal de l'Ecole Polytechnique, Vol. 58, 1889, p. 80.

[16] Rayleigh, J., "Aerial plane waves of finite amplitude." Proc. Roy. Soc., Vol. 84, 1910, pp. 274 - 284; Scientific Papers, Vol. 5, pp. 573 - 610.

[17] Love, A. E. H., and Pidduck, F. B., "Lagrange's ballistic problem." Trans. Roy. Soc. of London, Vol. 222, 1922, pp. 167 - 226. (Study of the motion during the expansion of high-pressure gas resulting from an explosion between the two free pistons in a cylindrical tube.)

[18] Becker, R., "Stosswelle und Detonation." Zeitschrift für Physik, Vol. 8, 1922. (In this fundamental paper the phenomenon of shock is investigated, with emphasis on physical aspects. The occurrences within the shock front are discussed in great detail. A large part of the paper is devoted to the problem of detonation. Ample historical references concerning experiments and theory are given throughout.)

[19] von Neumann, J., "Progress report on the theory of shock waves." NDRC, Division 8, OSRD No. 1140, 1943. Restricted. (Comprehensive report on the principles of compressible fluid flow, in particular on shocks. The notions of shock interaction, contact surface, shock reflection, and Mach effect are introduced. Further remarks on the process of detonation are included. Various reports of Division 8 of the NDRO, to which the authors had no access, are quoted here.)

For description and analysis of experimental work on shock waves, see, for example:

[20] Cranz, C., and Schardin, H., "Kinematographie auf ruhendem Film und mit extrem hoher Bildfrequenz." Zeitschrift für Physik, Vol. 56, 1929, pp. 147 - 183.

[21] Hilton, W. F., "The photography of airscrew sound waves." Proc. Roy. Soc., Series A, Vol. 169, 1938, pp. 174 - 190.

[22] Lewy, H., "On the relation between the velocity of a shock wave and the width of the light gap it leaves on the photographic plate." Aberdeen Proving Ground, Ballistic Research Laboratory, Report No. 373, 1943. Restricted.

[23] Keenan, Philip C., "Shadowgraph determination of shock-wave strength." Navy Department (Re2c), Explosives Research Report No. 11, 1943. Restricted. (See also references listed here.)

[24] Reynolds, G. T., "A preliminary study of plane shock waves formed by bursting diaphragms in a tube." NDRC, Division 2, Armor and Ordnance Report No. A-192 (OSRD No. 1519), 1943. Confidential.

For the analysis of interactions of one-dimensional shock waves with walls, with each other and with other types of waves, and for decaying shock waves, see [19] and

[25] Taylor, G. I., "The propagation and decay of blast waves." Home Office (A.R.P.D.), Civil Defence Research Committee, R. C. 39. (W-4-1). Confidential.

[26] Chandrasekhar, S., "On the decay of plane shock waves." Aberdeen Proving Ground, Ballistic Research Laboratory, Report No. 423, 1943. Restricted.

[27] Courant, R., and Friedrichs, K. O., "Interaction of shock and rarefaction waves in one-dimensional motion." NDRC, AMP Report 38.1R (AMG-NYU No. 1), 1943. Restricted.

[28] Cohn, H ., "Repeated reflection of a shock against a rigid wall." NDRC, AMP Memo 38.2M (AMG-NYU No. 3), 1943. Restricted.

[29] Chandrasekhar, S., "The normal reflection of a blast wave." Aberdeen Proving Ground, Ballistic Research Laboratory, Report No. 439, 1943. Restricted.

[30] von Neumann, J., "Proposal and analysis of a new numerical method for the treatment of hydrodynamical shock problems." NDRC, AMP Report 108.1R (AMG-IAS No. 1), 1944. Confidential.

For the theory of shock waves in gases with more general equations of state, see [19] and

[31] Weyl, H., "A scheme for the computation of shock waves in gases and fluids." NDRO, AMP Memo 38.7M (AMG-NYU No. 18), 1943. Confidential.

[32] Weyl, H ., "Shock waves in arbitrary fluids." NDRC, AMP Note No. 12 (AMG-NYU No. 46), 1944. Restricted.

For the theory of elastic-plastic waves (App. 2, Chapter III), see a great number of reports and memoranda by v. Kármán and others issued through Division 2 of the NDRC. For App. 3 see

[33] von Kármán, "Eine praktische Anwendung der Analogie zwischen Überschallströmung in Gasen und überkritischer Strömung in offenen Gerinnen." Zeitschrift für angewandte Mathematik und Mechanik, Vol. 18, No. 1, 1938, pp. 49 - 56, and references listed there.

For a discussion of shock and rarefaction waves in two-dimensional fluid flow, see [2] to [11] and

[34] Meyer, Th., "Über zweidimensionale Bewegungsvorgänge in einem Gas, das mit Ueberschallgeschwindigkeit strömt." (Dissertation, Göttingen, 1908.) Forschungsheft 62 des Vereins deutscher Ingenieure, Berlin, 1908, pp. 31 - 67. (This is the first paper which deals with oblique shock and rarefaction waves, following suggestions by Prandtl. Flow through nozzles is also discussed.)

[35] Epstein, P. S., "On the air resistance of projectiles." Proc. Nat. Acad. Sci., Vol. 17, 1931, pp. 532 - 547. (A treatment of the two-dimensional flow around a projectile of polygonal shape employing shock and rarefaction waves.)

[36] Schubert, F., "Zur Theorie des stationären Verdichtungsstosses." Zeitschrift für angewandte Mathematik und Mechanik, Vol. 23, No. 3, 1943, pp. 129 - 138.

For the theory of regular reflection see [19] and

[37] von Neumann, J., "Oblique reflection of shocks." Navy Dept., Bureau of Ordnance, Explosives Research Report No. 12 (Re2c), 1943. Confidential.

[38] Polachek, H., and Seeger, R. J., "Regular reflection of shocks in ideal gases." Navy Dept., Bureau of Ordnance, Explosives Research Report No. 13, 1944. Confidential. (Complete numerical data for various values of γ.)

[39] Keenan, P. C., and Seeger, R. J., "Analysis of data on shock intersections. Progress Report 1." Navy Dept., Bureau of Ordnance, Explosives Research Report No. 15, 1944. Confidential. (Includes a comparison of experimental and theoretical data for the Mach effect.)

The theory of the Mach effect is discussed in [19], [37] and

[40] Chandrasekhar, S., "On the conditions for the existence of three shock waves." Aberdeen Proving Ground, Ballistic Research Laboratory, Report No. 367, 1943. Restricted.

[41] Friedrichs, K. O., "Remarks on the Mach effect." and "Appendix to remarks on the Mach effect." NDRC, AMP Memos 38.4M and 38.5M, (AMG-NYU Nos. 5 and 6), 1943. Confidential.

[42] Friedrichs, K. O., "On configurations involving three shocks through one point modified by Meyer waves." NDRC, AMP Memo 38.9M (AMG-NYU No. 21), 1943. Confidential.

For theoretical studies of flow through nozzles and jets see [2], [3], [4] and

[43] Chaplygin, S. A., "On gas jets." Scientific Annals of the Imperial University of Moscow, Physico-Mathematical Division; Pub. No. 21, Moscow, 1904. Translation by Maurice H. Slud, available through the School of Mechanics, Brown University, Providence, R. I. (This basic paper deals with subsonic jets. It is remarkable that the

author, in common with Lamb and others, rejects the possibility of supersonic steady flow, although such flow had long been discovered and treated by engineers.)

[44] Stodola, A., Dampf- und Gasturbinen. 6th edition, Berlin, 1924. (Steam and Gas Turbines, translated by L. C. Lowenstein, McGraw-Hill, 1927.)

[45] Prandtl, L., and Busemann, A., "Näherungsverfahren zur zeichnerischen Ermittlung von ebenen Strömungen mit Überschallgeschwindigkeit." Stodola Festschrift, Zürich and Leipzig, 1929, pp. 499 - 509.

[46] Stanton, T. E., "On the flow of gases at high speeds." Proc. Roy. Soc. of London, Series A, Vol. 111, 1926, pp. 306 - 339.

[47] Busemann, A., "Lavaldüsen für gleichmässige Überschallströmungen." Zeitschrift des Vereins deutscher Ingenieure, Vol. 84, 1940, pp. 857 - 862.

[48] Kisenko, M. S., "Comparative results of tests on several different types of nozzles." Central Aero-Hydrodynamical Institute, Moscow, Report No. 478, 1940. (Translation, NACA, Technical Memorandum No. 1066, 1944.) (Papers by Frankl and others are quoted here.)

[49] "Theoretical studies on the flow through nozzles and related problems." NDRC, AMP Report 82.1R (AMG-NYU No. 43), 1944. Confidential.

[50] Emmons, Howard W., "The numerical solution of compressible fluid flow problems." National Advisory Committee for Aeronautics, Technical Note No. 932, 1944. Restricted.

[51] Prandtl, L., "Vorträge und Diskussionen von der 78. Naturforscherversammlung zu Stuttgart." Physikalische Zeitschrift, Vol. VIII, 1907, pp. 23 - 32.

[52] Hartmann, Jul., and Lazarus, F., "The air-jet with a velocity exceeding that of sound." Phil. Mag., Ser. 7, Vol. 31, 1941.

[53] Fraser, R. P., "Flow through nozzles at supersonic speeds." Four interim reports on jet research, July 1940 to June 1941. Ministry of Supply, D. S. R. Extra Mural Research F 72/115. (WA-1513-1a).

[54] Friedrichs, K. O., "Remarks about R. P. Fraser's reports on jet research." NDRC, AMP Memo 82.3M (AMG-NYU No. 47), 1944. Restricted.

For applications of the theory of jets and nozzles to rockets, reference may be made to:

[55] Malina, F. J., "Characteristics of the rocket motor unit based on the theory of perfect gases." Journal of Franklin Institute, Vol. 230, 1940, pp. 433 - 454.

[56] "Rocket Fundamentals." NDRC, Div. 3, Section H, OSRD No. 3711 (ABL-SR1), 1944. Confidential.

For theoretical analysis of supersonic flow about a cone and conical shocks see:

[57] Taylor, G. I., and Maccoll, J. W., "The air pressure on a cone moving at high speeds." Proc. Roy. Soc., Series A, Vol. 139, 1933, pp. 278 - 311.

[58] Maccoll, J. W., "The conical shock wave formed by a cone moving at high speed." Proc. Roy. Soc., Series A, Vol. 159, 1937, pp. 459 - 472.

[59] Busemann, A., "Die achsensymmetrische kegelige Überschallströmung." Luftfahrtforschung, Vol. 19, No. 4, 1942, pp. 137 - 144. Also earlier papers by Bourquard and others quoted there.

[60] Friedrichs, K. O., "Remarks on the reflection of conical shocks." NDRC, AMP Memo 38.8M (AMG-NYU No. 19), 1943. Confidential.

For analyses of axially symmetric and spherically symmetric supersonic flows, steady and non-steady, see:

[61] Taylor, G. I., and Davies, R. M., "Positive and negative impulses in blast waves of small intensity." Ministry of Home Security, R. C. 315, S. W. 8 (WA-211-6), 1942. Restricted.

[62] Taylor, G. I., "Summary of work on the physics of explosions, Part 2." Ministry of Home Security, R. C. 190a, 1941. Confidential.

[63] Taylor, G. I., "The formation of a blast wave by a very intense explosion." Ministry of Home Security, R. C. 210, (II-5-153), 1941. Confidential.

[64] Guderley, G., "Starke kugelige and zylindrische Verdichtungsstösse in der Nähe des Kugelmittelpunktes bzw. der Zylinderachse." Luftfahrtforschung, Vol. 19, No. 9, 1942, pp. 302 - 312. ("Powerful spherical and cylindrical compression shocks in the neighborhood of the center of the sphere and the cylinder axis." RTP Translation No. 1448, by M. Flint, issued by the Ministry of Aircraft Production in London.)

[65] Kirkwood, J. G., and Bethe, H. A., Progress report on "The pressure wave produced by an underwater explosion. I". NDRC, Div. B, Serial No. 252 (OSRD No. 588), 1942.

An extensive field of research involving compressible fluid flow has not been considered in the present book, viz., the flow of compressible fluids around obstacles, in particular airfoils. For a survey and an exhaustive bibliography, see:

[66] von Kármán, Th., "Compressibility effects in aerodynamics." Journal of the Aeronautical Sciences, Vol. 8, No. 9, 1941, pp. 337 - 356.

AUTHORS LISTED IN BIBLIOGRAPHY

LIST OF SYMBOLS

Numbers refer to pages on which symbols are defined.*

* Note that the same symbol may have different connotations, depending on where it is used.

CPSIA information can be obtained
at www.ICGtesting.com
Printed in the USA
LVHW05s1542020518
575706LV00015B/1050/P

9 781333 57330